SUPERCRITICAL FLUID
CHROMATOGRAPHY

RSC Chromatography Monographs

Series Editor: Roger M. Smith, *University of Technology, Loughborough, UK.*

Advisory Panel: J. C. Berridge (*Sandwich, UK*), G. B. Cox (*Delaware, USA*), I. S. Lurie (*Virginia, USA*), P. J. Schoenmakers (*Eindhoven, The Netherlands*), C. F. Simpson (*London, UK*), G. G. Wallace (*Wollongong, Australia*).

This series is designed for the individual practising chromatographer, providing guidance and advice on a wide range of chromatographic techniques with the emphasis on important practical aspects of the subject.

Supercritical Fluid Chromatography
Edited by Roger M. Smith, University of Technology, Loughborough, UK.

Forthcoming Titles

Chromatographic Integration Methods
by N. Dyson, *Dyson Instruments Ltd., Houghton le Spring, UK.*

Sample Preparation for HPLC
by R. D. McDowell, *Smith, Kline and French, Welwyn, UK.* and U. A. Th. Brinkman, *Amsterdam, The Netherlands.*

How to obtain future titles on publication
A standing order plan is available for this series. A standing order will bring delivery of each new volume immediately upon publication, at a substantial discount price. For further information, please write to:
  The Royal Society of Chemistry
  Distribution Centre
  Blackhorse Road
  Letchworth
  Herts. SG6 1HN

Telephone: Letchworth (0462) 672555

RSC
CHROMATOGRAPHY
MONOGRAPHS

Supercritical Fluid Chromatography

Roger M. Smith
Department of Chemistry
University of Technology
Loughborough

ROYAL
SOCIETY OF
CHEMISTRY

CHEMISTRY

02844059

British Library Cataloguing in Publication Data

Smith, Roger M. (Roger Malcolm)
Supercritical Fluid Chromatography—
    (RSC Chromatography Monographs)
    1. Liquid chromatography
    I. Title   II. Royal Society of Chemistry
    543′.0894   QD 79.C454

ISBN 0-85186-577-1

Published by The Royal Society of Chemistry
Burlington House, Piccadilly, London W1V 0BN

Typeset by Bath Typesetting Ltd., Bath,
and printed by St. Edmundsbury Press, Bury St. Edmunds, England

Preface

Although a chromatographic separation using a supercritical fluid as the mobile phase was first reported in 1962, supercritical fluid chromatography (SFC) has only attracted serious attention in the last few years, following the introduction of high efficiency open-tubular columns and availability of commercial instrumentation. As an analytical or preparative technique, it offers an alternative separation selectivity and different sampling requirements to the older established methods of GLC and HPLC and these differences have aroused considerable interest in the technique by analytical chemists, particularly in the food, petroleum, and pharmaceutical industries. However, information on the methods and capabilities of SFC has been primarily restricted to review articles and lectures. While these have spurred greater interest they have been necessarily restricted in their coverage.

The increasing knowledge about the methods, instrumentation, and applications of SFC suggested that it would be an appropriate topic to inaugurate the new RSC Chromatography Monograph Series being planned by the Royal Society of Chemistry. This proposal coincided with a Short Course in Supercritical Fluid Chromatography, which was organised in July 1986 by the Editor as part of the comprehensive series of training courses in Analytical Chemistry run by the Analytical Research Group at Loughborough University. This Course brought together speakers from Europe, USA, and Japan and also provided the opportunity for practical demonstrations on working instruments and for extensive informal discussions between participants, lecturers, and instrument manufacturers. A number of the lecturers responded to a invitation to contribute Chapters for this Monograph, and these have been supplemented by invited papers from other researchers in the field. All the contributions were prepared in the Summer of 1987.

Generally the papers concentrate on the methods employed in SFC separations and the ability of the chromatographer to control the separation. It has often been suggested that SFC offers the best of both GLC and HPLC, but while this is often an exaggeration, in practice SFC can provide a wide range of operating conditions, from gas-like separations similar to GLC to liquid-like separations more comparable to HPLC, with particular advantages for involatile and thermally labile analytes. At the extremes of the ranges, the divisions from the older chromatographic methods are often blurred and much of the instrumentation used in SFC has been based on modifications or adaptations of existing GLC or HPLC equipment and detectors and has gained from experience in these areas. The technique has great flexibility to alter retentions by adjusting or programming the temperature, pressure, and/or density, by altering the nature of the stationary phase, and by changing the properties of the mobile supercritical fluid phase by the addition of modifiers.

The applications and scope of SFC are still being explored and new areas and analytes are continually being reported. It has shown some special advantages for the coupling of SFC to mass spectrometry and these are discussed in some detail. The combination with supercritical fluid extraction as a novel sample preparation technique for SFC has been described but is yet to be fully explored and the ready removal of the mobile phase also suggests the potential of SFC for preparative separations.

Clearly, much is still to be learnt about the role of SFC as an analytical or preparative technique in the laboratory and its relationship to the established methods of HPLC and GLC and it is hoped that the present Monograph, by bringing together the different aspects of SFC, will help to promote further progress and will make available to a wider audience a greater understanding of the technique and its potentials and capabilities.

Roger M. Smith
Loughborough
October 1987

Contributors

Keith D. Bartle, *Department of Physical Chemistry, University of Leeds, Leeds*

Antony J. Berry, *Department of Chemistry, University College, P.O. Box 78, Cardiff, CF1 1XL*

D. E. Games, *Department of Chemistry, University College, P.O. Box 78, Cardiff, CF1 1XL*

T. Hondo, *Jasco Spectroscopic Co., Tokyo, Japan*

Stephen J. Lane, *Glaxo Group Research, Greenford, Middlesex*

Dietger Leyendecker, *Dionex Corp., Richard-Klinger-Str 15, 6270 Idstein, West Germany*

I. C. Mylchreest, *Department of Chemistry, University College, P.O. Box 78, Cardiff, CF1 1XL*

J. R. Perkins, *Department of Chemistry, University College, P.O. Box 78, Cardiff, CF1 1XL*

S. Pleasance, *Department of Chemistry, University College, P.O. Box 78, Cardiff, CF1 1XL*

Muneo Saito, *Jasco Spectroscopic Co., Tokyo, Japan*

M. Marsin Sanagi, *Department of Chemistry, University of Technology, Loughborough, Leicestershire, LE11 3TU*

Pat Sandra, *Research Institute for Chromatography, Belgium*

Roger M. Smith, *Department of Chemistry, University of Technology, Loughborough, Leicestershire, LE11 3TU*

Peter J. Schoenmakers, *Phillips Research, Eindhoven, The Netherlands*

Y. Yamauchi, *Jasco Spectroscopic Co., Tokyo, Japan*

Contents

Glossary of Chromatographic Terms

d_c	Internal diameter of open-tubular column
d_f	Film thickness of the stationary phase
d_p	Particle size of stationary phase
D_M	Diffusion coefficient in the mobile phase
D_S	Diffusion coefficient in the stationary phase
D_{11}	Self diffusion coefficient of the mobile phase
D_{12}	Binary diffusion coefficient in the mobile phase
h	Height equivalent to a theoretical plate (HETP)
h_r	Reduced plate height h/d_c (open-tubular columns) or h/d_p (packed columns)
k'	Capacity factor
K	Distribution constant
L	Column length
n	Column efficiency, number of theoretical plates
n_{ne}	Required number of theoretical plates
N	Column efficiency, number of effective theoretical plates
P	Pressure
P_C	Critical pressure
P_R	Reduced pressure (P/P_C)
P_V	Vapour pressure at 0.7 T_C
R_s	Resolution of two peaks.
S_a	Specific surface area
T	Temperature
T_C	Critical temperature
T_R	Reduced temperature (T/T_C)
t_0	Column void volume
t_M	Mobile phase retention time
t_R	Retention time from injection
t'_R	Adjusted retention time ($t_R - t_0$)
$\bar{u}$	Average linear velocity of mobile phase (L/t_0)
u	Linear velocity of mobile phase (L/t_0)
u_{opt}	Optimum flow rate for minimum h
V_M	Volume of mobile phase in column
V_S	Volume of stationary phase in column
w_b	Width of peak at base (4 σ)
w_h	Width of peak at half height (2.35 σ)
z	Compressibility factor
α	Phase ratio V_S/V_M

α	Separation ratio, relative capacity factors or relative adjusted retention times (k_2'/k_1')
δ	Solubility parameter
δ_f	Reduced film thickness $d_f/d_c\sqrt{D_M/D_S}$
η	Viscosity
λ	Eddy diffusion coefficient
ν	Reduced velocity $\bar{u}d_c/D_M$ (open-tubular columns) or $\bar{u}d_p/D_M$ (packed columns)
φ	Solvent fugacity constant
ρ	Density

CHAPTER 1

Theory and Principles of Supercritical Fluid Chromatography

KEITH D. BARTLE

1 Introduction

Although chromatography with a supercritical fluid as mobile phase was reported more than twenty years ago, the advantages of supercritical fluid chromatography (SFC) have only recently been realised—in particular its rapidity, flexibility, and ability to allow the analysis of substances which cannot be analysed by gas chromatography (GC). The separating power of open-tubular column GC is unparalleled, but its applicability is restricted by the limited volatility and thermal stability of many organic compounds; less volatile compounds can be analysed by high performance liquid chromato-graphy (HPLC), but long analysis times and very small column diameters are required for efficient separations because of the limitations of solute diffusion in the mobile phase. SFC overcomes these difficulties and permits high resolution at low temperatures with short analysis times.

Above its critical point a substance such as carbon dioxide has properties which make its use as a chromatographic mobile phase very favourable. The ability of supercritical fluids to dissolve substances was first recorded in 1879, when Hannay and Hogarth[1] studied the solubility of cobalt and iron chlorides in supercritical ethanol. Numerous applications of extraction with supercritical solvents have been described, and the underlying theory fully discussed.

Following a suggestion by Lovelock[2] in 1958 that a supercritical fluid might be used as a mobile phase in chromatography, Klesper *et al.* first demonstrated[3] SFC by the separation of nickel porphyrins using super-critical chlorofluoromethanes as mobile phases. Sie and Rijnders[4] and Giddings[5] developed the technique further, both practically and theoretically, and many applications have been reported. An important further develop-ment was the demonstration of SFC with capillary columns by Novotny and

Lee[6] in 1981. Landmarks in the history of SFC are listed in Table 1. The use of SFC in analysis is now widespread, and is increasing apparently exponentially. A number of commercial instruments are on the market.

Table 1 *Landmarks in Supercritical Fluid Chromatography*

Event	Researchers	Date
Demonstration of solubility in supercritical fluids	Hannay and Hogarth	1879[1]
First suggestion of SFC	Lovelock	1958[2]
Separation of porphyrins by SFC	Klesper, Corwin and Turner	1962[3]
Developments of SFC	{ Sie and Rijnders	1967[4]
	{ Giddings	1966[5]
Analysis of petroleum derived mixtures by SFC	Jentoft and Gouw	1972
Capillary SFC	Novotny and Lee *et al.*	1981[6]
Commercial packed-column SFC instrument		1981
Commercial capillary column SFC instrument		1985

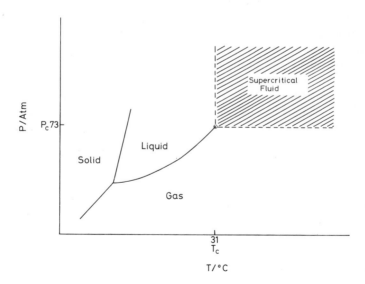

Figure 1 *Carbon dioxide phase diagram*

2 Supercritical Fluids

A supercritical fluid may be defined from a phase diagram (Figure 1), in which the regions corresponding to the solid, liquid, and gaseous states are clear. Above the critical temperature, however, the vapour and liquid have the same density, and the fluid cannot be liquefied by increasing the pressure. The supercritical region in the P-T diagram (Figure 1) is indicated by dashed lines since no phase change occurs. The definition of a supercritical fluid is therefore arbitrary, in that there is a continuous transition from liquid to supercritical fluid by increasing the temperature at constant pressure (through the vertical line in Figure 1) or from gas to supercritical fluid by increasing the pressure at constant temperature (through the horizontal dashed line in Figure 1). Supercritical fluids are now widely used in extraction, fractionation, and chromatography.[7,8,9]

Table 2 *Physical Parameters of Selected Supercritical Fluids*

Fluid	Critical Temperature $T_C(°C)$	Critical Pressure $P_C(atm)$	Critical Density $\rho_C g/ml^{-1}$	Density at 400 atm. $\rho_{400\,atm}$ g/ml^{-1}	Density of liquid $\rho_L g/ml^{-1}$
CO_2	31.3	72.9	0.47	0.96	0.93 (63.4 atm, 25 °C)
N_2O	36.5	72.5	0.45	0.94	0.91 (sat., 0 °C)
					0.64 (59 atm, 25 °C)
NH_3	132.5	112.5	0.24	0.40	0.68 (sat., −33.7 °C)
					0.60 (10.5 atm, 25 °C)
$n-C_5$	196.6	33.3	0.23	0.51	0.75 (1 atm, 25 °C)
$n-C_4$	152.0	37.5	0.23	0.50	0.58 (sat., 20 °C)
					0.57 (2.6 atm, 25 °C)
SF_6	45.5	37.1	0.74	1.61	1.91 (sat., −50 °C)
Xe	16.6	58.4	1.10	2.30	3.08 (sat., 111.75 °C)
CCl_2F_2	111.8	40.7	0.56	1.12	1.53 (sat., −45.6 °C)
					1.30 (6.7 atm, 25 °C)
CHF_3	25.9	46.9	0.52	—	1.51 (sat., −100 °C)

Some suitable mobile phases for SFC are listed in Table 2. In addition to their critical temperatures and pressures, the chromatographically important properties of supercritical fluids are the density, viscosity, and the diffusion coefficients of solutes.[9] Above its critical point, a substance has density and solvating power approaching that of a liquid, but viscosity similar to that of a gas, and diffusivity intermediate between those of a gas and liquid (Table 3). Hence, the supercritical fluids in Table 2 have properties which make their use as chromatographic phases very favourable. As long as intermolecular interactions are sufficiently strong, supercritical fluids are able to dissolve a variety of solutes, even those with high molecular mass and low volatility. The density of supercritical fluids, and hence the solubility and chromatographic retention of solutes, can easily be changed by changing the applied pressure.

Table 3 *Comparison of Chromatographic Techniques*[17]

Mobile Phase	Density (g ml^{-1})	Viscosity (poise)	Diffusivity (cm^2 s^{-1})	Column Types
Gas	$\approx 10^{-3}$	0.5–$3.5(\times 10^{-4})$	0.01–1.0	capillary
Supercritical fluid	0.2–0.9	0.2–$1.0(\times 10^{-3})$	3.3–$0.1(\times 10^{-4})$	capillary and packed
Liquid	0.8–1.0	0.3–$2.4(\times 10^{-2})$	0.5–$2.0(\times 10^{-5})$	packed

The two other important properties of supercritical fluids are the viscosity and solute diffusion coefficients. The low viscosity means that the pressure drop across the column for a given flow rate is greatly reduced. Since the viscosities of gases and supercritical fluids are about the same, but liquid viscosities are greater by a factor of approximately 100, the pressure drop in HPLC is between 10 and 100 times greater than in SFC and GC. The greater permeability of capillary columns, of course, allows column length to be increased. The mass transfer properties resulting from solute diffusion coefficients in supercritical fluids lead to analysis speeds which increase in the sequence HPLC, SFC, and GC.

3 Supercritical Fluid Chromatography

The instrumentation necessary for SFC[10,11] is illustrated in Figure 2. The mobile phase is pumped as a liquid, and the pressurised fluid is preheated above the critical temperature before passing into the column *via* an injection valve, and hence into the detector. A pressure restrictor is located either after the detector or at the end of the column to ensure supercritical conditions.

Packed-column SFC can be used to chromatograph thermally labile compounds and compounds beyond the volatility range of GC; its chief advantage over HPLC is a reduction in analysis time. The longer column lengths possible in capillary SFC give the advantage of high efficiency and allow the analysis of complex mixtures which cannot be separated by mobile phase selectivity.

In contrast to HPLC pumping systems, pressure rather than flow control is necessary, and pulseless operation is more critical. Syringe pumps are most commonly employed, therefore. Provision must also be made for pressure or density programming—slowly increasing the mobile phase density to decrease solute retention. This is the analogue of temperature programming in GC and gradient elution in HPLC. However, the addition of small quantities of polar solvents as modifiers can also radically alter the retention behaviour of analytes. The oven temperature must be held constant during a run to prevent variation in the density of the mobile phase which may affect retention.

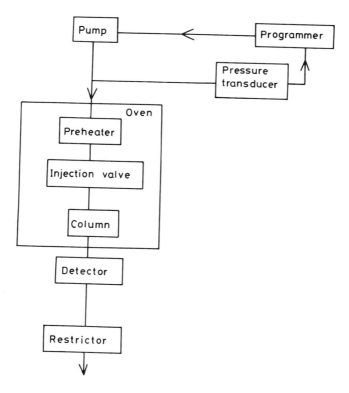

Figure 2 *Block diagram of a supercritical fluid chromatograph*

The small column diameters necessary in capillary SFC make the method of sample introduction important. Small volumes and rapid injection are necessary. SFC is compatible with both GC and HPLC detectors.[12] Optical detectors (UV absorption and fluorescence) must have very small volumes and be able to withstand high pressures. Fluid decompression and expansion in the flame jet must be accommodated in flame detection (flame ionization, thermionic, and flame photometric detectors). Promising developments have been recorded in coupled systems such as SFC-mass spectrometry[13] and SFC-Fourier transform infrared spectroscopy.[14,15]

4 A Comparison of Chromatographic Techniques

The basic equation describing separation in chromatography defines the resolution, R_s, between any two compounds. In the chromatogram, R_s is derived from the peak maxima separation, Δt, and the baseline peak width, w_{b1} and w_{b2} (Figure 3):

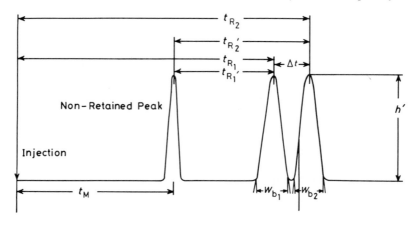

$$R_S = \frac{\Delta t}{[w_{b1} + w_{b2}]/2}$$

$R_S \geqslant 1$ *Satisfactory resolution*
$R_S = 1.5$ *Baseline resolution*

Figure 3 *Resolution in chromatography*

$$R_s = \frac{2\Delta t}{(w_{b1} + w_{b2})} \tag{1}$$

Resolution is satisfactory for $R_s \geqslant 1$; 'baseline' (99.9%) resolution corresponds to $R_s = 1.5$. In terms of the column efficiency, n, the selectivity, α, and the capacity factor, k', R_s is given by:

$$R_s = \frac{n^{\frac{1}{2}}}{4} \cdot \frac{(\alpha - 1)}{\alpha} \cdot \frac{k'}{(1 + k')} \tag{2}$$

where, in terms of retention times:

$$n = 16(t_R/w_b)^2 \tag{3}$$

$$k' = t_R - t_M/t_M \tag{4}$$

(t_M is the retention time of an unretained substance)

$$\text{and } \alpha = k_2'/k_1' \tag{5}$$

The larger the selectivity or the efficiency, of course, the greater the separation. The number of theoretical plates necessary to achieve a given resolution at a given k' may be calculated from a rearranged version[16] of equation (2):

$$n = 16R_s^2 \left(\frac{\alpha}{\alpha - 1} \right)^2 . \frac{(k' + 1)^2}{k'} \tag{6}$$

Figure 4 is a graph of n against α for $R_s = 1.5$ and $k' = 2$, *i.e.* baseline separation with a peak with retention twice that of the holdup volume.[16] For HPLC with packed columns, $n \approx 5 - 10 \times 10^3$ plates, so that the minimum selectivity is 1.10; for capillary GC $n \approx 1 - 2 \times 10^5$ plates and the selectivity is ≈ 1.03. These differences are a reflection of the $\approx 10^5$ lesser diffusion coefficients of solutes, D_M, in the liquid mobile phase as compared with diffusion coefficients in gases. For SFC, values of D_M are approximately 10^2 times greater than in liquids, and separating power midway between GC and HPLC is expected. Representative chromatograms[17] for a complex mixture are shown in Figure 5: in fact the separation achieved by SFC on a 34 m × 50 µm i.d. capillary column falls little short of that obtained by GC with the same stationary phase on a 20 m × 300 µm i.d. capillary. Smaller internal diameters are necessary to allow for the reduced rates of diffusion in supercritical mobile phases.

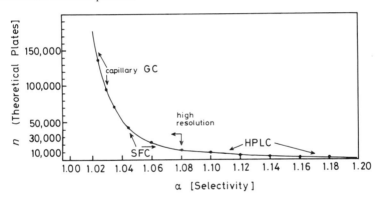

Figure 4 *Graph of number of theoretical plates,* n, *required to achieve a separation with resolution 1.5 at* k' = 2, *against selectivity,* α *(Equation 6)*
(Reproduced by permission from Gere, Board, and McManigill, Pittsburgh Conference on Analytical Chemistry, Atlantic City, NJ, March, 1982.)[16]

Further comparisons of the different chromatographic methods may be made[18] with the aid of the van Deemter equation, which relates h, the column 'height' (or length) corresponding to a theoretical plate, to average linear mobile phase velocity u:

$$h = A + \frac{B}{u} + C_M u + C_S u \tag{7}$$

where the first and second terms are, respectively, flow uniformity and longitudinal diffusion terms, and the third and fourth express resistance to mass transfer in mobile and stationary phases respectively.

(a)

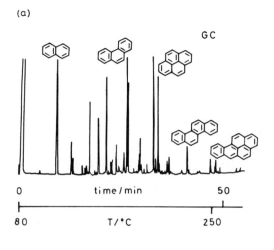

GC

```
0          time / min          50
```

```
80            T/°C            250
```

(b)

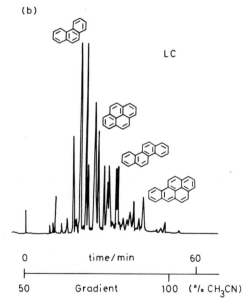

LC

```
0          time/min          60
```

```
50         Gradient      100   (°/° CH₃CN)
```

Figure 5 (a) *and* **(b)** *Comparison of chromatographic techniques: separation of coal tar by GC, and HPLC*
(M. L. Lee, Private communication, 1986.)[17]

For packed columns, assuming no resistance to mass transfer in the thin liquid film[19]:

$$h = 2\lambda d_{\mathrm{p}} + \frac{2D_{\mathrm{M}}}{u} + \frac{d_{\mathrm{p}}^2(1 + 6k' + 11k'^2)u}{24D_{\mathrm{M}}(1 + k')^2} \tag{8}$$

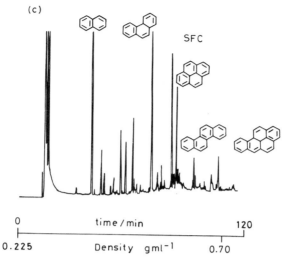

Figure 5 (c) *Comparison of chromatographic techniques: Separation of coal tar by SFC*
(M. L. Lee, Private communication, 1986.)[17]

where d_p is the column packing particle diameter and λ is the eddy diffusion coefficient. Thus, for SFC and HPLC experiments on the same packed column under conditions giving similar values of k' for the same solute, van Deemter plots of h against $\bar{u}$ show (Figure 6) similar minimum plate heights, h_{min}, since this is a function of d_p and k' only:

$$h_{min} = 2d_p\left[\lambda + \frac{(1 + k' + 11k'^2)^{\frac{1}{2}}}{\sqrt{12}(1 + k')}\right] \qquad (9)$$

The optimum linear velocity u (*i.e.* value of u corresponding to h_{min}):

$$u_{opt} = 4\sqrt{3D_M}\Big/ \frac{d_p(1 + 6k' + 11k'^2)^{\frac{1}{2}}}{(1 + k')} \qquad (10)$$

depends directly on D_M, however, and is hence higher in SFC, thus reducing the analysis time for a given resolution.

For capillary columns, the A term in the van Deemter equation is zero, and the internal diameter of the column, d_c, replaces the particle diameter. The h *versus* u curves calculated as above for SFC and GC on the same capillary column with representative values of D_M at the same k' again yield the same h_{min} since this depends only on k' and d_c; the dependence of u_{opt} on D_M now greatly increases the value of u_{opt} for the GC mode (Figure 6).

The much reduced column operating temperatures in SFC compared with GC represent an enormous advantage in applications to high molecular mass and thermolabile substances, and allows 'brute force' high-resolution chromatography to be applied to mixtures which normally would be inaccessible to GC. Such mixtures could be separated by HPLC but only after considerable investment of time to determine the appropriate mobile phase composition. The increased resolution in unit time possible in SFC is a further advantage over HPLC.

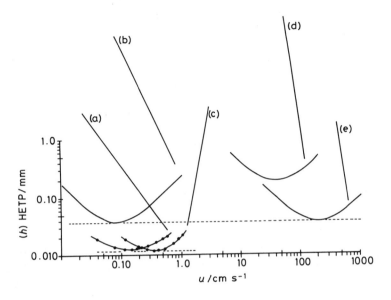

Figure 6 *Van Deemter plots for GC, HPLC and SFC*
    (a) *HPLC: ODS*, 5μm, 10 cm × 4.6 mm, $k' = 2.85$, CH_3CN/H_2O
    (b) *Capillary SFC:* $d_c = 50\,\mu m$, $k' = 2.30$, CO_2, 0.8 g ml^{-1}, 40 °C, $D_{12} = 0.00008\,cm^2\,s^{-1}$
    (c) *SFC: ODS*, 5 μm, 10 cm × 4.6 mm, $k' = 2.30$, CO_2, 0.8 g ml^{-1}, 40 °C
    (d) *GC:* $d_c = 250\,\mu m$, $k' = 2.30$, $D_{12} \approx 0.2\,cm^2\,s^{-1}$
    (e) *GC:* $d_c = 50\,\mu m$, $k' = 2.30$, $D_{12} \approx 0.2\,cm^2\,s^{-1}$
    (Reproduced by permission from Randall, *A.C.S. Symp. Series No.* **250**, 1984, p. 135.)[19]

5 Operating Parameters in Supercritical Fluid Chromatography

The Mobile Phase in SFC

The partition coefficient, and hence the value of k' for a solute in SFC, is altered by varying the density of the mobile phase. In turn, the variation of density with pressure (see Chapter 3, Figure 5) is required and this is calculated by means of reduced parameters, as follows. The true molar volume at a specified temperature and pressure is calculated from the ideal gas law and the compressibility factor, z, at reduced temperature and pressure, T_R and P_R:

$$z(T_R,P_R) = z'(T_R,P_R) + wz''(T_R,P_R) \qquad (11)$$

w is the acentric factor

$$w = -\log(P_V/P_C) - 1 \qquad (12)$$

where P_V is the vapour pressure at 0.7 times the critical temperature, and P_C is the critical pressure. Values of z' and z'' have been tabulated[20] as a function of T_R and P_R.

Software programmes are available to calculate densities as a function of pressure at a given temperature and to fit these data to an nth order polynomial. Programmed elution by linear or asymptotic density profile is then achieved by changing the pressure so as to produce the desired density according to the above polynomial.

To carry out in SFC the analogue of temperature programming in GC or gradient elution in HPLC, the density, ρ, must be increased during the run so as to elute regularly members of a homologous series, carbon number n.[21] The density programming rate may be predicted from the expression:

$$\ln k' = A_0 + B_0 n - mn\rho \qquad (13)$$

where A_0, B_0 and m are constants. From equation (13) it can be shown[21] that:

$$\rho = \frac{B_0}{m} - \frac{1}{mF'n} \qquad (14)$$

where $B_0/m = \rho_A$ is the threshold density at which all members of the homologous series co-elute. It follows that resolution decreases as ρ approaches ρ_A, and in linear density programming successive oligomers are eluted closer together, as is clear from Figure 7.

If the members of the series are to be eluted at regular time intervals, t:

$$n = j(t + t') \qquad (15)$$

where j is a constant and t' is the reference elution time.
Hence:

$$\rho = \frac{(A_0 - \ln k')}{mj(t + t')} + \frac{B_0}{m} \qquad (16)$$

Combining variables:

$$\rho = \rho_A - \frac{K}{(t + t')} \qquad (17)$$

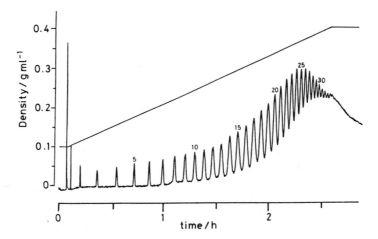

Figure 7 *Separation by SFC of a mixture of polystyrene oligomers with a linear density programme*
(Reproduced by permission from *J. Chromatogr. Sci.*, 1983, **21**, 222.)[21]

Asymptotic density programming according to equation (17) is possible if ρ_A is known and, as Figure 8 shows, allows elution of oligomers with regular spacing. There are two convenient methods for obtaining values of ρ_A: firstly, by making constant-density runs and plotting ln α *vs.* ρ, the density corresponding to $\rho = 1$ is determined. Secondly, if the elution density for a series of oligomers is plotted against $1/n$ in a linear density run, ρ_A is found by extrapolating to $1/n = 0$.

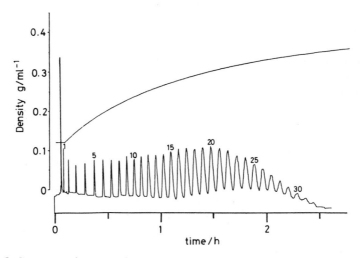

Figure 8 *Separation by SFC of a mixture of polystyrene oligomers with an asymptotic density programme*
(Reproduced by permission from *J. Chromatogr. Sci.*, 1983, **21**, 222.)[21]

As density increases, the diffusion coefficient of a solute decreases and the plate height increases, while the lower D_M for high molecular mass solutes also increases the plate height. Both of these effects lead to significant losses in efficiency in capillary column SFC at high densities (see Section 6).

Mobile Phase Composition in SFC

Density programming in SFC is not possible at the highest densities and the addition of small quantities of polar modifiers may be necessary to migrate certain solutes. In packed column SFC, small ($<1\%$) amounts of modifier produce large retention changes, presumably because the modifier competes with the solute for active sites in the packed column.[18,19] For capillary columns, however, larger quantities of modifier (5–20% in CO_2) are required[22] to achieve any marked effect on retention (Table 4). Comparison of modifier effects for equivalent densities should be made, since mobile phase density is the main factor controlling isothermal retention, but in the absence of acccurate data, similar temperature and pressure conditions must suffice as a basis for the comparison.

Table 4 *Effects of Polar Modifiers in CO_2 on Retention*[a][22]

SOLUTE	k'			α		
	CO_2	IPA[b]/CO_2	CH_3NO_2/CO_2	CO_2	IPA/CO_2	CH_3NO_2/CO_2
carbazole	2.2	0.4	0.2	0.2	1.4	1.4
pyrene	4.4	0.6	0.3	2.1	1.2	—
4-hydroxy pyrene	9.1	0.8	—	—	—	—

[a] Conditions: 7.5 mole % modifier in CO_2 at 80 °C, 125 atm for i-propanol and 135 atm for CH_3NO_2
[b] IPA = i-propanol

Modifiers for CO_2 may be selected by an approach based on that of Snyder for HPLC solvents.[23] Briefly, the contributions of proton donor, proton acceptor, and dipole interactions to total polarity are plotted on a triangular diagram (Figure 9); the total polarity of each solvent is indicated on the Figure. The data allow the first-order classification of solvents according to their position in the Snyder triangle.[22]

For mixed mobile phases, the critical constants of the mixture, T_C and P_C, can be approximated[24] as the arithmetic mean of the critical temperatures T_A and T_B and the critical pressures P_A and P_B, *i.e.*

$$T_C = x_A T_A + x_B T_B$$

$$P_C = x_A P_A + x_B P_B$$

where x_A and x_B are the mole fractions of components A and B. More

Theory and Principles of SFC

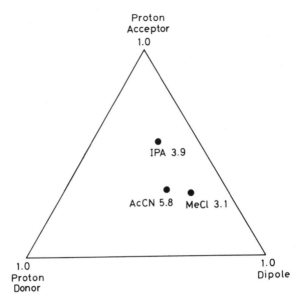

Figure 9 *Snyder triangle diagram of modifier solvent interactions. Total polarity is listed beside each modifier*
(Reproduced by permission from Fields, Thesis, Brigham Young University, Provo, Utah, 1986.)[22]

elaborate treatments are based on the method of Cheuh and Prausnitz[25] for T_C, and Kreglewski and Kay[26] for P_C.

To employ density programming in SFC with mixed mobile phases, the dependence of density on pressure is required. An empirical procedure has been proposed[27] in which the density (ρ_i) is calculated by determining the dead time (t_L) of the unmodified phase as a liquid at 20 °C (ρ_L) and then the dead time (t_i) of the (modified) supercritical fluid of interest at pressures near the critical pressure and temperatures above the critical temperature

$$\rho_i = \rho_L \, (t_i/t_L) \, (P_i/P_L) \tag{18}$$

where P_i is the pressure drop ($P_{inlet} - P_{outlet}$) for the modified phase, and P_L the pressure drop for the unmodified liquid. Equation (18) follows from the Poiseuille equation if the viscosity of the mixed fluid at room temperature is close to the viscosity of the unmodified liquid. Densities are determined at a range of pressures and fitted to a sixth order polynomial.

Columns for Supercritical Fluid Chromatography

The strong solvating ability of mobile phases in SFC makes the careful selection of stationary phases imperative. Much early work employed absorbents such as alumina, silica, or polystyrene, or stationary phases insoluble in supercritical CO_2 or pentane such as polyethylene glycol. More recent packed-column work has involved bonded non-extractable stationary phases such as octadecylsilyl and aminopropyl bonded silicas.

From Equation (8) it follows that the plate height depends on the square of particle diameter; reducing the particle size reduces the HETP so that columns packed with 3 μm particles are preferable to 5 or 10 μm particles. However, the high surface area of small-diameter column packings may bring about catalytic interaction with reactive solutes. The advent of new column technology[28] and the injection and detection technology necessary for their utilization have led to increasing use of wall-coated capillary columns in SFC. These have the overriding advantage of high plate numbers because of greater length consequent on greater permeability, but are also valuable in requiring low mobile phase flow rates—thus permitting the easier use of toxic, corrosive, inflammable, or expensive fluids. Capillary columns are also compatible with a variety of detectors, permit maximum use of density programming, and allow high sensitivity from narrow peaks. The small pressure drops across capillary columns result in only small selectivity changes. The stationary phase, usually a polysiloxane—methyl-, phenyl-methyl-, octyl-, biphenyl-, and cyanopropylsiloxanes have proved particularly useful—must be subjected to one of the free radical cross-linking methods now standard for capillary GC columns so that the stationary phase film is not extracted by the mobile phase.[28]

The relative speed of analysis of packed and column SFC may be compared[29] by means of the parameter h_{min}/u_{opt}. For $1 < k' < 5$:

$$h_{min}/u_{opt} = \frac{2}{3} \times \frac{d_p^2}{D_M} \text{ for packed columns} \qquad (19)$$

$$h_{min}/u_{opt} = 0.1 \frac{d_c^2}{D_M} \text{ for capillary columns} \qquad (20)$$

For equal speeds of analysis, therefore, $d_c = 2.6d_p$. For a given separation, a 50 μm i.d. capillary column corresponds to a column packed with ≈ 20 μm particles. In this respect, the readily available 3 and 5 μm particles correspond to 9 and 13 μm i.d. capillaries, so far beyond the capabilities of column preparation technology. However, the smaller pressure drop across the capillary column allows longer columns with larger plate numbers. For equal pressure drops:[30]

$$n_c/n_p = 4.6 \, (d_c/d_p)^2 \qquad (21)$$

where n_c and n_p are the maximum number of theoretical plates on capillary and packed columns respectively. A 50 μm capillary column therefore generates more than 100 times as many plates as a column packed with 10 μm particles and more than 450 times as many as one packed with 5 μm particles.

The full van Deemter equation for capillary columns (the Golay equation) is:

$$h = \frac{2D_M}{u} + \frac{d_c^2(1 + 6k' + 11k'^2)u}{96D_M(1 + k')^2} + \frac{2k'd_f^2u}{3(1 + k')^2D_S} \tag{22}$$

where d_f is the stationary phase film thickness. The term in d_c^2 in Equation (22) ensures a decrease in HETP and an increase in the number of plates per unit length as the internal diameter of the SFC capillary column decreases. This effect has been confirmed[31] for van Deemter plots, both calculated (with the aid of literature diffusion coefficients) and experimental. Values of u_{opt} calculated from Equation (22) are inversely proportional to d_c, increasing from 0.1 to 0.4 cm s^{-1} as the column decreases from 100 to 25 μm, for $k' = 3$. Practical efficiencies obtained in a chromatographic run may be illustrated by plotting plates per metre against k' (Figure 10). A 50 μm i.d. column produced 4500–3800 plates m^{-1} for $k' = 4–11$. For $u = 2$ cm s^{-1} at $k' = 1.35$, the 50 μm i.d. column produced 7500 plates m^{-1}. Even at $20\ u_{opt}$ ($u = 4.3$ cm s^{-1}) the 50 μm column produced 3700 plates m^{-1} at $k' = 5$.

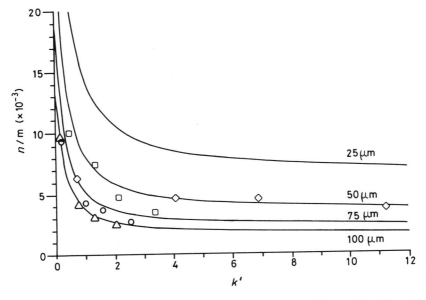

Figure 10 *Graph of efficiency per unit length against* k' *for different column diameters.* CO_2 *at 40 °C, 72 atm, 0.22 g ml^{-1}, 10 u_{opt}* *(Reproduced by permission from J. High Resolut. Chromatogr., Chromatogr. Commun., 1984, 7, 312.)*[31]

The stationary phase film thickness, d_f, contribution to plate height in capillary SFC, $C_s u$, was evaluated theoretically[32] with the same diffusion coefficients as were used for the evaluation of the effect of column diameter.

For solutes with $k' = 1-5$ on 50 μm i.d. columns with CO_2 mobile phase there was no significant difference in h_{min} or u_{opt} as d_f was increased from 0.25 to 1.0 μm. These results were then confirmed experimentally. Figures 11 and 12 show, respectively, how van Deemter curves and graphs of plates per metre against k' vary little for d_f up to 1 μm.

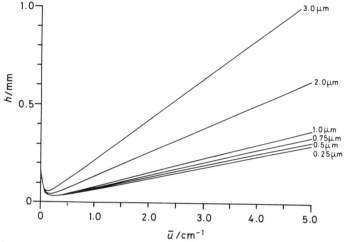

Figure 11 *Van Deemter plots for different film thicknesses.* CO_2 *at* 40 °C *and* 72 atm, 50 μm *i.d. columns coated with* SE–54, $k' = 1$
(Reproduced by permission from *J. High Resolut. Chromatogr., Chromatogr. Commun.*, 1984, **7**, 423.)[32]

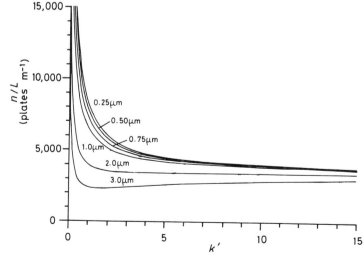

Figure 12 *Graph of efficiency per unit length against* k' *for different film thicknesses.* CO_2 *at* 40 °C *and* 72 atm, 2.0 cm s^{-1} *linear velocity*, 50 μm *i.d. columns coated with* SE–54
(Reproduced by permission from *J. High Resolut. Chromatogr., Chromatogr. Commun.*, 1984, **7**, 423.)[32]

Great demands are placed on the sensitivity of detectors in SFC by the small diameter of the columns necessary to obtain high efficiency. One approach is to increase the stationary phase film thickness and hence the sample capacity. This increases the ratio of stationary phase plate-height contribution ($C_S u$) to mobile phase plate height ($C_M u$). The ratio $C_S u/C_M u$ is proportional to the ratio of mobile phase to stationary phase solute diffusion coefficients. Since, as noted in Section 2, D_M/D_S in GC is $\approx 10^4$ but only $\approx 10^2$ in SFC, the same increase in film thickness under SFC conditions would not be expected to result in as large an increase in plate height as would be found under GC conditions.

The preparation of small-diameter capillary columns for SFC[28] is less straightforward than for GC columns with the more usual internal diameter of 200–500 μm. The Rayleigh instability of films of stationary phase solution during coating[33] involves the fourth power of column radius, a; the time τ, for a disturbance to increase in amplitude from b_0 to b is given by:

$$\tau = \ln \frac{b}{b_0} \times \frac{12\eta a^4}{\gamma d_f{}^3} \tag{23}$$

where η and γ are, respectively, the viscosity and surface tension of the coating solution and d_f is the thickness of the film. It follows that a film on the wall of a 50 μm i.d. column will be 256 times less stable than a corresponding film on 200 μm i.d. column. Modifications to established coating procedures are being developed, however, to allow the preparation of such columns.

Injector and Detector Requirements

Injection in SFC is usually achieved by means of the switching of the contents of a sample loop into the carrier fluid at the column entrance by means of a suitable valve. For packed column SFC a conventional HPLC injection system is adequate, but for capillary column work facility is generally provided for the injected volume to be introduced in split or splitless mode. The very small injector volumes required in capillary SFC may be determined[29] from the general equation relating the effects of a sample plug entering or leaving a column on observed peak widths:

$$\frac{\tau_i}{\tau_c} = \frac{n_i}{n} \left(\frac{n}{12} \right)^{\frac{1}{2}} \tag{24}$$

τ_i and τ_c are respectively the standard deviations in the peak widths due to the injector (or detector) and the column; n and n_i are, respectively, the numbers of theoretical plates provided by the column and the number of those plates 'filled' by the injector or detector volume. Combining equations (1)–(4) and (24) allows an expression to be derived[29] linking fractional loss of resolution, ΔR_s, to V_i, the injector or detector volume to column and retention characteristics:

$$V_i = 0.866\,\pi d_c^2 (Lh)^{\frac{1}{2}} [1/(1 - \Delta R_s)^2 - 1]^{\frac{1}{2}} (1 + k') \tag{25}$$

since h is a function of d_c, detector volumes are proportional to $d_c^{\frac{5}{2}}$; column length must be increased by a factor of four to double permissible volumes.

The small volumes which would cause only a 1% loss in resolution are shown in Table 5. For the commonly used 50 µm i.d. columns, loop volumes of 50–100 nl are therefore used, and these must be switched in precise time intervals; pneumatically driven valves are employed.

On-column detection in capillary SFC (UV absorption, fluorescence, *etc.*) is feasible even for 50 µm i.d. columns; column lengths of a few cm (L_i in Table 5) may be used. If, as is usual, sample application and detection takes place outside the column oven, *i.e.* the eluent will be in the liquid state rather than the supercritical state, the values of V_i and L_i in Table 5 must be divided by the ratio of densities of liquid and supercritical fluid.

Table 5 *Injector or Detector Volumes Resulting in a 1% Loss in Resolution*[a]

Column i.d.	Injector/detector volume	Column length
$d_c(\mu m)$	$V_i(\mu l)$	$L_i(cm)$
200	1.5	4.8
100	0.27	3.4
50	0.05	2.4

[a] Calculated from Equation (25) for 20 m long column, plate height $= 0.6d_c$, $k' = 1$ [29]

6 Resolution in Supercritical Fluid Chromatography

The equation defining the resolution of a chromatographic column as a function of efficiency, selectivity, and capacity factor has been given above, Equation (2). In considering resolving power in SFC it is important to take into account the role of pressure through its effect on retention, selectivity, and diffusion.[29] The pressure drop across the column in SFC is approximately linear, and if the density change with pressure is also linear then:

$$\rho = \rho_{inlet} - wL \tag{26}$$

where L is the column length and w is a constant. It follows that the average fluid density is the average of the inlet and outlet densities.

A more accurate average density may be calculated by integrating with respect to pressure (a function of viscosity) across the whole length of column. With the aid of an appropriate equation of state, calculations have been carried out[34] for CO_2 which show that, at inlet pressures above 150 bar at 40 °C with pressure drops below 30 bar, the deviation from the simple average value is less than 1%. With increasing pressures and temperatures away from the critical point, the deviation falls correspondingly. Only near the critical point is the deviation significant.

The effect of mobile phase density on selectivity is given by:[29]

$$\ln\alpha = B_0 - m\rho \tag{27}$$

while for a given solute a simpler form of Equation (13) can be written:

$$\ln k' = a - b\rho \tag{28}$$

where a and b are constants. The diffusion coefficient in the mobile phase is related to both viscosity, η, and the density of the mobile phase:

$$D_m = Z/\eta\rho \tag{29}$$

where Z is a constant.

Equations (26) and (28) can be combined to obtain an expression for k', which after integration over column length and dividing by the column length yields the observed k' value:

$$k'_{obs} = e^{(a - b\rho[\text{inlet}])}(e^{bwL} - 1)/bwL \tag{30}$$

Similarly, the observed selectivity is obtained from Equations (26) and (27)

$$\alpha_{obs} = e^{(B - m\rho[\text{inlet}])}(e^{mwL} - 1)/mwL \tag{31}$$

Since the density drop across the column, $\Delta\rho$, is simply wL, it follows that changes in observed selectivity and retention values depend only on the total density drop.

The effect of pressure drop on column efficiency is small.[29] This comes about because the product of D_M and ρ is almost constant [Equation (29)]. The effect of pressure drop on resolution, through its effect on selectivity and retention, has been investigated by determining the values of the constants in Equations (27) and (28) for substitution into Equations (30) and (31). A number of conclusions were drawn:

(a) α_{obs} changes by ≈ 0.001 unit per 10% density drop up to 30%;
(b) the sensitivity of α_{obs} to density drop increases at lower k' values;
(c) when α_{obs} is large, density drops have little effect on resolution;
(d) selectivity decreases more rapidly as density drops become larger;
(e) selectivity increases at larger k' where the mobile phase density is lower.

Table 6 *Pressure Drop Effects on Resolution at Maximum Density Change with Pressure*[29]

α	$\Delta R_s = -5\%$ $(\Delta\rho/\rho_{\text{inlet}})\%$	$(\Delta P/P)\%$	$(\Delta P/P)\% = 10\%$ Loss in $R_s\%$
1.005	1.3	0.3	100
1.02	5.1	1.3	40
1.05	13.1	3.3	15

The percentage density and pressure drop giving 5% resolution losses are shown in Table 6 at various α, in addition to the percentage resolution lost with 10% pressure drop. The loss in selectivity is dramatic at larger pressure drops (Figure 13), and this makes small-particle diameter packed columns less useful in the pressure-temperature region where the change in density is greatest. For capillary columns, however, the pressure drop is small; calculations based on the Poiseuille equation suggest that less than 3% drop at $10\,u_{\text{opt}}$ is expected for 50 μm columns of conventional length even at low k' (Figure 14). Clearly resolution losses from pressure drops across capillary columns are small.

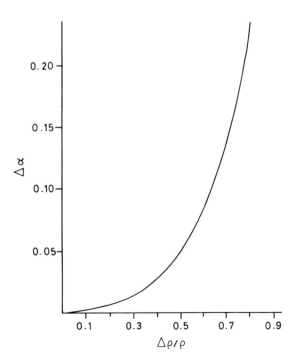

Figure 13 *Change in selectivity, $\Delta\alpha$, resulting from fractional density drop across column, $\Delta\rho/\rho$*
(Reproduced by permission from *J. Chromatogr.* 1983, **259**, 1.)[29]

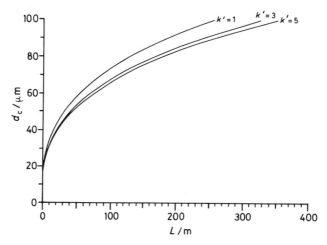

Figure 14 *Relation between column diameter,* d_c, *and length,* L *for a 3% pressure drop at different* k' CO_2 *at 40 °C, 72 atm, 0.22 g ml^{-1}, 10 u_{opt} (Reproduced by permission from J. High Resolut. Chromatogr., Chromatogr. Commun., 1984, 7, 312.)[31]*

The maximum resolution limit in SFC may be defined[29] as the smallest α which may be resolved, and is limited by the number of theoretical plates and the loss of selectivity $\Delta\alpha$, resulting from pressure drop: at the limit, $\Delta\alpha$ is equal to $1-\alpha$ and, from the above considerations is approximately $-0.02\Delta\rho/\rho_{inlet}$, which is near $-0.08\Delta\rho/\rho$ in the region where density changes most rapidly with pressure (see Chapter 3, Figure 5). Values of $\Delta\rho$ may then be obtained from the Poiseuille equation and incorporated into Equation (2) to yield:

$$0 = \left[1 - \frac{4R_s}{n^{\frac{1}{2}}}\left(\frac{1+k'}{k'}\right)\right]^{-1} - 2.56\frac{\eta unh}{d_c^2 P} - 1 \qquad (32)$$

Substitution of appropriate values in Equation (32) ($k' = 5, D_M = 2 \times 10^{-4}$ cm^2 s^{-1}, $D_s = 1 \times 10^{-6}$ cm^2 s^{-1}, $\eta = 5 \times 10^{-4}$ g cm^{-1} s^{-1}, $P = 40$ bar, $u = 10\, u_{opt}$) allows calculation of n by successive iterations. The use of 50 μm i.d. columns generates $>10^5$ effective theoretical plates in approximately 2 hours (Table 7).

Table 7 *Maximum Performance for Various Column Diameters*[29]

d_c(μm)	α_{min}	L(m)	u(cm s^{-1})	N	h(mm)	$\Delta P/P$	t_r(days)
200	1.008	856	0.60	9.7×10^5	0.61	0.05	9.9
100	1.013	169	1.20	3.9×10^5	0.30	0.08	0.98
50	1.020	34	2.40	1.6×10^5	0.15	0.13	(2.3 hr)

Table 8 *Calculated Effect of Density and More-retained Solutes on Efficiency*[22]

u (cm s^{-1})	ρ (g ml^{-1})	Solute	h (mm)	multiple of h_{min}	n/L m^{-1}
2	0.28	benzene	0.103	2.9	9,700
	0.45	naphthalene	0.167	4.6	6,000
	0.79	caffeine	0.413	11.5	2,400
1	0.28	benzene	0.057	1.6	17,500
	0.45	naphthalene	0.086	2.4	11,600
	0.79	caffeine	0.208	5.8	4,800

At high mobile phase densities, for example towards the end of a density programme required to elute a highly retained solute, the D_M of a solute decreases with a consequent increase in plate height. More retained solutes are, in any case, expected to have smaller D_M. The influence of both of these effects on column efficiency may be calculated with the aid of literature values of D_M, and examples are listed in Table 8. The efficiencies during a density-programmed run could decrease to about a quarter.

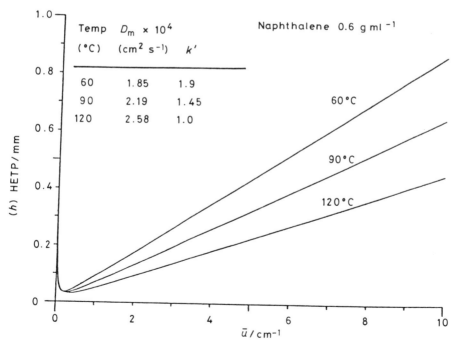

Figure 15 *Effect of temperature on efficiency: van Deemter plots at different temperatures*
(Reproduced by permission from Fields, Thesis, Brigham Young University. Provo, Utah, 1986.)[22]

The loss of efficiency at higher mobile phase densities may be partly overcome[35] by increasing the operating temperature at constant density where plate heights are significantly reduced (Figure 15). As noted in Section 7, k' is reduced, while D_M is increased. However, at higher temperatures, u and plate height increase more rapidly with density, thus offsetting the effects described above. Care must be taken, moreover, not to increase the temperature beyond the limit of stability of a thermolabile solute. Higher temperature operation requires higher pressure to attain high densities, and the pressure limits of column materials must not be exceeded.

7 Retention in Supercritical Fluid Chromatography

Plots of k' against temperature in SFC are characteristically shaped with a pronounced maximum above the critical temperature. Qualitatively this may be explained[36] by an increase of the free volume in the mobile phase which leads to a reduction in solubility and a shift in partition in favour of the stationary phase. On increasing the temperature, the vapour pressure and solubility of the solute increases and the concentration in the stationary phase decreases and is shifted towards the mobile phase, hence reducing k'. The variation of k' with temperature is more usually demonstrated *via* plots of log k' *vs.* $1/T$, where T is absolute temperature, at constant pressure (Figure 16) which also show the 'turnover' effect. Plots of log k' *vs.* $1/T$ at constant density are, however, linear, even as the temperature passes through the critical temperature. Values of k' are reduced and migration rates are increased by increasing the temperature at constant density; the increase in solubility of solutes in supercritical fluids with increasing temperature at constant density was pointed out by Hannay in 1880.[37]

The variation of log k' with $1/T$ for homologues is shown in Figure 16. The vertical distance between the curves at a given $1/T$ is, of course, log α, from Equation (5), and it appears that little variation in selectivity between homologues is possible by changing the temperature.[38] For dissimilar compounds, however, the log k' *vs.* $1/T$ curves may cross, representing a reversal in elution sequence.

Chester and Innis[38] have applied a thermodynamic approach to explain the variation of log k' with $1/T$ in SFC. In Figure 17 the region of positive slope corresponds to GC-like behaviour, whilst the region of negative slope is more like LC. These combinations may be represented by:

$$\log k' = \frac{-0.43\Delta H_s}{RT} - \log \beta + \frac{0.43\Delta H_m}{RT} \tag{33}$$

where ΔH_s and ΔH_m are, respectively, the partial molar heats of solution of the solute in the stationary phase and mobile phases respectively, and β is the column phase ratio. ΔH_m is a function of the mobile phase density, since, as

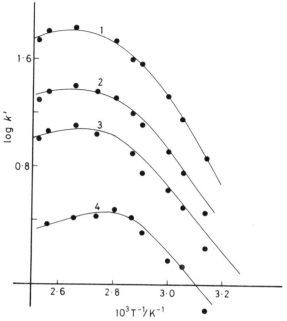

Figure 16 *Graph of logarithm of retention parameter,* k', *against reciprocal of absolute temperature at constant pressure of* CO_2 *for:* 1. *pyrene;* 2. *phenanthrene;* 3. *fluorene; and* 4. *naphthalene. Column* 25 cm *ODS*

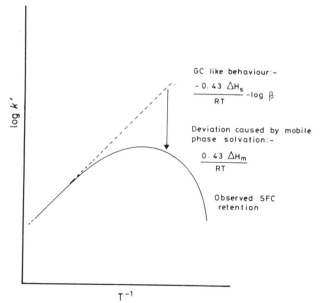

Figure 17 *Model of observed SFC retention behaviour*
(Reproduced by permission from *J. High Resolut. Chromatogr., Chromatogr. Commun.*, 1985, **8**, 561.)[38]

expected from Equation (28), plots of log k' against average density at constant temperature are linear. It may be concluded from the treatment in Equation (33) that the enthalpy change in removing solute from the stationary phase to the mobile phase under GC conditions (*i.e.* vaporization) is reduced by the enthalpy of solvation by the mobile phase.

A different thermodynamic approach[39] can be applied to account for the variation of log k' with $1/T$ in SFC. The solute fugacity coefficient, φ, is employed so that graphs of log $k' - $ log φ *versus* $1/T$ (*e.g.* Figure 18) are linear. From the modified van der Waals equation of state of Peng and Robinson,[40] φ is calculated, in which the attractive pressure term is modified so as to be more accurate at high fluid phase densities.

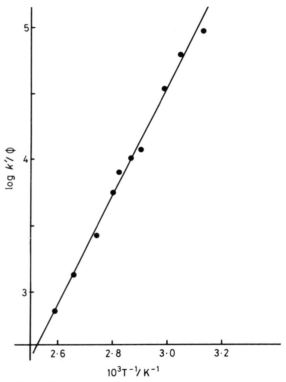

Figure 18 *Graph of logarithm of retention parameter divided by fugacity coefficient against reciprocal of absolute temperature at constant pressure of* CO_2 *for fluorene. Column:* 25 cm *ODS*

8 Conclusions

SFC offers significant advantages over both HPLC and GC. Both GC and HPLC detectors may be employed, and the use of capillary columns in SFC leads to efficiencies comparable with GC. All the four significant operating parameters in chromatography are varied in SFC (Table 9) to achieve variations in retention and selectivity, so that the possibilities for multidimensional SFC are much increased. The low temperatures at which even higher molecular weight compounds are eluted by the solvating power of supercritical fluids lead to improvements in selectivity (so that for example, chiral recognition is more likely) and, most important, allows high resolution chromatography on thermally labile compounds.

Table 9 *A Comparison of Chromatographic Variables*

	GC	HPLC	SFC
Temperature	+	−	+
Mobile phase composition	−	+	+
Stationary phase	+	+	+
Pressure	−	−	+
High resolution	+	−	+

+ Parameter which can be varied readily to alter retention and selectivity

Acknowledgements

This work was supported by the Science and Engineering Research Council, British Gas p.l.c., The Royal Society of Chemistry, and the British Council. Many helpful discussions with my colleagues Ian K. Barker, Anthony A. Clifford, Jacob P. Kithinji, Mark W. Raynor and Gavin F. Shilstone, and especially with Milton L. Lee are gratefully acknowledged.

References

1. J. B. Hannay and J. Hogarth, *Proc. Roy. Soc. (London)*, 1879, **29**, 324.
2. J. Lovelock, 1958, private communication quoted in W. Bertsch, Thesis, University of Houston, Texas, 1973.
3. E. Klesper, A. H. Corwin, and D. A. Turner, *J. Org. Chem.*, 1962, **27**, 700.
4. S. T. Sie and G. W. A. Rijnders, *Sep. Sci.*, 1966, **1**, 469; *ibid.*, 1967, **2**, 699, 729, 755.
5. J. C. Giddings, *Sep. Sci.*, 1966, **1**, 73.
6. M. Novotny, S. R. Springston, P. A. Peaden, J. C. Fjeldsted, and M. L. Lee, *Anal. Chem.* 1981, **53**, 407A.
7. P. F. M. Paul and W. S. Wise, 'The Principles of Gas Extraction', Mills and Boon, London, 1971.
8. D. F. Williams, *Chem. Eng. Sci.*, 1981, **36**, 1769.
9. P. A. Peaden and M. L. Lee, *J. Liq. Chromatogr.*, 1982, **5**(Suppl. 2), 179.
10. K. D. Bartle, I. K. Barker, A. A. Clifford, J. P. Kithinji, M. W. Raynor, and G. F. Shilstone, *Anal. Proc.*, 1987, **24**, 299.
11. J. C. Fjeldsted and M. L. Lee, *Anal. Chem.*, 1984, **56**, 619A.
12. M. Novotny, *J. High Resolut. Chromatogr.*, *Chromatogr. Commun.*, 1986, **9**, 137.
13. B. W. Wright, H. T. Kalinoski, H. R. Udseth, and R. D. Smith, *J. High Resolut. Chromatogr.*, *Chromatogr. Commun.*, 1986, **9**, 145.

14. M. W. Raynor, K. D. Bartle, I. L. Davies, A. Williams, A. A. Clifford, J. Chalmers, and B. Cook, *Anal. Chem.*, in the press.
15. S. L. Pentoney, K. H. Shafer, and P. R. Griffiths, *J. Chromatogr. Sci.*, 1986, **24**, 230.
16. D. Gere, R. Board, and D. McManigill, Pittsburgh Conference on Analytical Chemistry, Atlantic City, NJ, March 1982.
17. M. L. Lee, Private communication (1986).
18. D. Gere, *Science*, 1985, **222**, 258.
19. L. G. Randall *in* 'Ultra High Resolution Chromatography', *ed.* S. Ahuja, *A. C. S. Symp. Series No.* 250, Washington, D.C. 1984, p. 135.
20. J. C. Fjeldsted, Thesis, Brigham Young University, Provo, Utah, 1985.
21. J. C. Fjeldsted, W. P. Jackson, P. A. Peaden, and M. L. Lee, *J. Chromatogr. Sci.*, 1983, **21**, 222.
22. S. M. Fields, Thesis, Brigham Young University, Provo, Utah, 1986.
23. L. R. Snyder, *J. Chromatogr. Sci.*, 1978, **16**, 223.
24. R. C. Reed and T. K. Sherwood, 'Properties of Gases and Liquids', 2nd ed., McGraw-Hill, New York, 1966.
25. P. L. Cheuh and J. M. Prausnitz, *Am. Inst. Chem. Eng. J.*, 1967, **13**, 1099.
26. A. Kreglewski and W. B. Kay, *J. Phys. Chem.*, 1969, **73**, 3359.
27. W. P. Jackson, R. C. Kong, and M. L. Lee, *in* 'PAH–Mechanisms, Methods and Metabolism (8th International Symposium)', *ed.* M. Cooke and A. J. Dennis, Battelle Press, Columbus, 1985, p. 609.
28. B. A. Jones, K. E. Markides, J. S. Bradshaw, and M. L. Lee, *Chromatography Forum*, May–June 1986, p. 38.
29. P. A. Peaden and M. L. Lee, *J. Chromatogr*, 1983, **259**, 1.
30. P. J. Schoenmakers, *in* 'Proceedings of Eighth International Symposium on Capillary Chromatography', *ed.* P. Sandra, Huethig, Heidelberg, 1987, p. 1061.
31. S. M. Fields, R. C. Kong, J. C. Fjeldsted, M. L. Lee, and P. A. Peaden, *J. High Resolut. Chromatogr., Chromatogr. Commun.*, 1984, **7**, 312.
32. S. M. Fields, R. C. Kong, M. L. Lee, and P. A. Peaden, *J. High Resolut. Chromatogr., Chromatogr. Commun.*, 1984, **7**, 423.
33. K. D. Bartle, C. L. Woolley, K. E. Markides, M. L. Lee, and R. S. Hansen, *J. High. Resolut. Chromatogr., Chromatogr. Commun.*, 1986, **9**, 610.
34. G. F. Shilstone, K. D. Bartle, and A. A. Clifford, to be published.
35. S. M. Fields and M. L. Lee, *J. Chromatogr.* 1985, **349**, 305.
36. E. Klesper and D. Leyendecker, *Int. Lab.*, 1986, **16** (9) November, 18.
37. J. B. Hannay, *Proc. Roy. Soc. (London)*, 1880, **30**, 484.
38. T. L. Chester and D. P. Innis, *J. High Resolut. Chromatogr., Chromatogr. Commun.*, 1985, **8**, 561.
39. G. F. Shilstone, J. P. Kithinji, A. A. Clifford, I. K. Barker, and K. D. Bartle, to be published.
40. D. Y. Peng and D. B. Robinson, *Ind. Eng. Chem. Fundam.*, 1976, **15**, 59.

CHAPTER 2

The Emergence and Instrumentation of SFC

M. MARSIN SANAGI and ROGER M. SMITH

1 The Development of SFC

Although supercritical fluid chromatography (SFC) was first demonstrated 25 years ago, there has only been widespread interest in the technique in the last few years. This is demonstrated in the sudden and exponential increase in papers since the early 1980s (Figure 1), and a widening of their geographical origins. From 1962 to mid-1987 Chemical Abstracts has indexed over 350 papers on SFC. The numbers have risen from 9 papers in 1982 to 39 in 1984, and over 90 were published in 1986. Up to the middle of 1987 a further 50 papers have appeared confirming the upward trend, and in this year a number of conferences and courses specifically devoted to this technique took place.

As with many new analytical methods, for the first few years only a small group of adherents studied the background and theory of SFC and it wasn't until instrumental developments made the method more readily available and new applications prompted a wider role that the technique attracted greater interest.

The emergence of SFC can be attributed, to a large extent, to the continual search in chromatography and other separation techniques for increased separation power to facilitate the resolution of more and more complex mixtures. In the early 1980s HPLC using microbore chromatographic columns and then capillary columns, either packed or wall-coated, attracted widespread academic and commercial interest.[1-3] They appeared to hold the theoretical promise of a separation power in HPLC comparable to the highly successful and widely adopted open-tubular columns in GLC. However, despite many commercial claims, it was soon found that this promise could be achieved only with great difficulty, both in instrument design and detection sensitivity and with a considerable time penalty. Consequently, few laboratories have adopted these methods for routine studies. A more considered approach to the theory confirmed that these problems could be

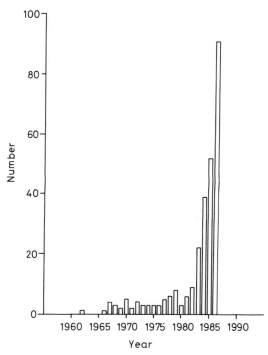

Figure 1 *Distribution of published articles on SFC (by year) from 1962 to 1986 based on a survey, July, 1987*

related to the properties of the liquid mobile phase, particularly its low diffusivity and high viscosity, coupled with the limited sensitivity of most HPLC detectors. In a search for a more suitable mobile phase, attention then turned to supercritical fluids.

In one of the most far reaching comments in any scientific paper, in 1941 Martin and Synge[4] proposed that their discussion of liquid chromatography would be generally applicable and 'the mobile phase may be a vapour' thus forecasting their eventual development of gas liquid chromatography. This acknowledgement that the mobile phase need not only be a liquid also implied the possibility of the use of supercritical fluids.

Although the solvating power of supercritical fluids has been known since 1879,[5] it was not until 1962 that Klesper, Corwin, and Turner published the first report of chromatography with a supercritical fluid as the mobile phase.[6] Using supercritical dichlorodifluoromethane or monochloro-difluoromethane, they demonstrated the separation of involatile nickel etioporphyrin II from nickel mesoporphyrin IX dimethylester. In 1970 Karayannis and Corwin extended this investigation with studies of the separation of a number of porphyrins and metal chelates.[7,8] Meanwhile, Sie and Rijnders were studying the use of carbon dioxide, isopropanol, and n-pentane as mobile phases with packed columns. They applied this technique, which they called 'fluid-solid chromatography (FSC)' or 'fluid-liquid chromatography (FLC)', to the analysis of polynuclear aromatic hydro-carbons and oxygenated compounds of coal tar.[9-12] At the same time, Giddings and his co-workers[13-18] further contributed to the early develop-ment of SFC through their studies on the separation of a variety of compounds, including carotenoids, sugars, nucleosides, amino acids, and polymers, using 'Ultra-high gas chromatography' or 'Dense gas chromato-graphy' with carbon dioxide or ammonia as the mobile phase.

In the 1970s, the development of SFC was relatively slow. This was probably because of a combination of factors, including the experimental problems in using supercritical fluids and a lack of commercial SFC instrumentation. Furthermore, the developments were largely overshadowed by the tremendous expansion of HPLC, following higher efficiencies obtained after the introduction of pellicular particles by Kirkland,[19] and the subsequent development of bonded and microparticular phases. However, the scope of SFC was being steadily broadened. Altares[20] and Jentoft and Gouw[21-24] applied SFC to the separation of monodisperse styrene oli-gomers and polynuclear aromatic compounds in automobile exhaust, while Nieman and Rogers[25] applied this technique to the characterisation of siloxane-based gas chromatographic stationary phases.

It wasn't until SFC was applied to open-tubular columns that a wider interest was aroused. In 1981 Novotny *et al.* reported the feasibility of the separation of non-volatile solutes with an efficiency comparable to capillary GLC, using a fused silica capillary column with n-pentane as the mobile phase.[26]

A number of patents on SFC instrumentation designs and technologies have been submitted and upheld since the early developments of SFC. Vitzthum, Huber, and Barthels from H.A.G. (Germany), an established company known for the application of supercritical fluid extraction methods in industry, patented a design in 1971 for a flame ionization detector for SFC.[27] Subsequently, in 1977 Hartmann and Klesper[28] patented the "Chromatographic separation of substances with low vapor pressures". Perrut, from Société Nationale Elf Aquitaine S.A. (France), patented a method and apparatus for fractionating mixtures by chromatography, with elution by a supercritical fluid,[29] while in Japan, Yokogawa-Hewlett Pack-ard patented an apparatus for concentrating gas and liquid chromato-graphic samples which was also applicable to SFC.[30] The following year, Morinaga Confectionary Co. Ltd. (Japan) patented a 'supercritical-fluid analyser', which was an apparatus for delivery of CO_2 mobile phase[31] and in 1985 Vidrine and Allhands (Nicolet Instrumentation Corp.) patented a light-pipe flow cell for SFC-FTIR.[32]

One of the most interesting patents in the history of SFC was filed for in early 1982 by Novotny, Lee, Peaden, Fjeldsted, and Springston for the fundamental use of open-tubular capillary columns for SFC.[33] This patent had aroused some controversy in the SFC industry, and consequently it was subjected to a review in late 1985 and 1986. On December 31, 1986 the claims of the patent were preliminarily rejected. However, after amendments were made, the fundamental patent for capillary SFC patent was validated in March, 1987. It is held by Brigham Young University and licensed exclusively to Lee Scientific[34] and these groups have done much to promote the subsequent interest in SFC.

2 The Literature of SFC

Publications and research into SFC have taken two main directions, the first concentrated in the USA and the latter more in Europe. The former body of work, reviewed later in this Chapter, has examined the equipment and methodology for SFC. This work has drawn heavily on systems used in GLC and HPLC, and particular interest has been focused on the use of different detectors and the possible role of SFC-mass spectrometry, which is considered in more detail in Chapters 6 and 7.

The second and, until recently, more minor area has been the understanding of the retention mechanism and the effect on retention and selectivity of changing the separation conditions and this aspect is discussed in detail in Chapters 3 and 4.

An examination of the contents of the papers on SFC published up to mid-1987 shows the way in which the technique has developed. The papers can be divided into four main areas: review, instrumentation, application, and theory (Table 1) and these will be considered in turn with the exception of the last. It was difficult to draw strictly distinctive criteria for the different types, and a paper may be grouped into one or any combination of the four different types.

Table 1 *Distribution of publications on SFC by type (1962–mid-1987)*

Type of paper	Number*
Review	75
Theoretical	88
Instrumentation	76
Application	133
SFE-SFC	11
Dissertation	13
Patent	7

* Based on 353 papers

Reviews on SFC

Since 1962, the authors of about half of the publications (excluding reviews) have described an application of SFC, while the instrumentation and theory have each been covered in about 30% of the publications. Surprisingly, nearly a quarter of all the publications of SFC have been reviews, containing few new results. These reviews, nonetheless, have played an important role in promoting the general interest in SFC. Together, they have served as an effective means of propagating the attractions in SFC by raising an awareness of SFC and conveying the new developments and applications of this technique to the scientific community. Their scope is surveyed below as they can provide an important source for further information.

One of the earliest reviews was by Giddings in 1968, entitled 'High pressure gas chromatography of non–volatile species'.[35] A few years later, Gouw and Jentoft[36] reviewed the general aspects of SFC, including different possible mobile phases, solute retention, selectivity, and applications. Schomburg[37] in his review the following year indicated the possibility of using 'fluid chromatography' as a successful separating method, although he was aware of the difficulties caused by the lack of suitable instrumentation, which hindered the application of the technique at that time. Huang[38] and Przyjazny *et al.*[39] reviewed techniques, apparatus, and applications of SFC. The effect of pressure and temperature, and the selection of phases were also discussed. Gouw and Jentoft in a further review[40] discussed the history, range of operation, and instrumentation aspects of SFC. Several reviews published in the 1970s[41–43] highlighted new developments in SFC, which were generally limited to packed-column techniques.

Parallel with the rapid development in capillary SFC in recent years, a number of reviews describing the general aspects of this technique have been published.[44–55] Novotny[47] reviewed the physical principles of SFC, and the general instrumentation, and described the possible use of several detection and ancillary techniques in capillary SFC. These include flame-based and plasma detectors, mass spectrometry and Fourier transform infrared spectroscopy. Richter[54] introduced the concept of capillary SFC, as well as the general instrumentation and applications. In another review, Caude and Rosset[48] discussed aspects of packed and capillary columns for SFC. Fjeldsted and Lee,[45] and Later *et al.*[49] in their reviews on capillary SFC included a discussion on the direction of SFC and the future areas of research. In a recent review, Lee and Markides[50] described the state-of-the-art of capillary SFC and discussed the progress that had been achieved in column technology.

The application of supercritical fluids has been exploited beyond SFC into the area of supercritical fluid extraction (SFE) and numerous review articles have described the combination of supercritical fluid extraction with SFC.[56–63] In 1982, Randall[58] presented a thorough review of work in the area of 'dense gas extraction' and 'dense gas chromatography' (some of which may be correctly referred to as SFC). In that article he also included a comprehensive list of work in these two areas giving details of the mobile phase, compounds separated, stationary phase, and detector.

The basis of SFC has been discussed by Schneider[63] and Van Wassen *et al.*[64] who reviewed the physicochemical principles of extraction and chromatography with supercritical fluids. Peaden and Lee[65] reviewed the principles of SFC, drawing on previous work to illustrate how variables such as temperature and pressure affect resolution. Klesper and Leyendecker[66] described retention and resolution in SFC using polynuclear aromatic hydrocarbons as test solutes. Yonker *et al.*[67] and Schneider[68,69] have discussed recent studies on physicochemical properties of supercritical fluid solutions which contributed to the better understanding of SFC.

Solute retention in SFC is governed by a number of factors, including the stationary and mobile phases, temperature, pressure/density, and mobile phase velocity. Because some of these parameters can be varied during a chromatographic separation, it is, therefore, possible to alter the retention and the selectivity by gradient methods. Schmitz and Klesper[70] reviewed solute retention behaviours in single eluents and gradient elution SFC. In another review[71] they discussed the practice of SFC using single and multiple gradients, including temperature, pressure, and velocity gradients.

Recent advances in the detection methods have made supercritical fluid chromatography an attractive and practical alternative for the separation and identification of complex mixtures. Novotny[72] reviewed the current status and development trends in detection in SFC. The use of GLC-like and HPLC-like detectors and ancillary identification methods were discussed. A number of review articles[52,53,58,73-77] discussed the use of SFC with mass spectrometric detection and the design of the SFC-MS interface. Smith, Wright, and Udseth have discussed the historical development, current instrumentation, applications, and future directions of capillary SFC-MS.[77] The coupling of Fourier transform infrared detectors has been reviewed by Griffith *et al.*[78] and more recently by Wieboldt and Adams[79] and Jinno.[80]

Numerous review articles[81-100] discussed the principles, methods, advantages, applications, and advances of SFC. Novotny[101,102] reviewed miniaturised separation systems including capillary SFC. The biochemical and environmental applications of the systems were emphasised. Levy[103] described the principles of the technique, operation, and performance of an instrument designed by Combustion Engineering with several examples of SFC separations including fatty acids and oil refinery side stream samples. In a recent article,[104] Novotny reviewed advances and trends in microcolumn LC and open-tubular SFC emphasising new detection and ancillary techniques. Greibrokk *et al.*[105] reviewed techniques and applications in SFC and discussed the possibilities of SFC compared to other chromatographic techniques. The limitations of SFC were also discussed, including the aspects of solute solubility, availability of SFC instrumentation, and injection techniques.

3 Instrumentation of SFC

Supercritical fluid chromatography instrumentation resembles both GC and HPLC equipment and because of this, it has been possible to take advantage of technological developments from both techniques. Capillary SFC was developed on principles based on capillary GC, while nearly all previous SFC instruments employed components normally used in conventional

HPLC systems, including high-pressure pumps, stainless steel tubing, injection valves, and columns, with few modifications, or none at all.

Separation Systems: Columns and Eluents

Much of the instrumentation in SFC is dependent on the column type, *i.e.* whether it is capillary or packed-column. Up to mid-1987, approximately 113 papers on open-tubular capillary column SFC, and 109 papers on packed-column SFC have been published. Caude and Rosset,[48] Schwartz,[106] and Schoenmakers[107] have discussed the advantages and limitations of packed and capillary SFC. More details are presented in Chapter 4 by Schoenmakers. In another report, Schwartz *et al.* discussed packed and capillary columns for SFC in terms of the separation efficiency and speed of analysis.[108] The most widely used stationary phases in packed columns for SFC were pure silica (26 papers), and octadecyl-bonded silica (17 papers). Other types of stationary phases used in packed-column SFC include amino, cyano, and octyl bonded silicas. Polysiloxane-based stationary phases usually bonded to the column wall, or cross-linked, have been predominant in capillary SFC, and SE-54 has been the most popular material.

Table 2 *Examples of frequently reported compounds used as mobile phase or organic modifiers in SFC*

Compound	Number of papers*
Acetonitrile	6
Ammonia	4
Carbon dioxide	121
1,4-Dioxane	7
Diethyl ether	3
Di-i-propyl ether	2
Hydrocarbons:	63
Pentane	39
Hexane	11
Butane	5
Ethane	4
Ethylene	2
Alcohols:	46
Methanol	30
Isopropanol	8
Nitrous oxide	7
Tetrahydrofuran	4

* Figures based on a survey, July, 1987.

A wide range of compounds can be converted to supercritical fluids and the potential of many have been examined as eluents and modifiers in SFC (Table 2). Carbon dioxide has been the single most widely used mobile phase. n-Pentane came next, followed by hexane. Other common eluents include nitrous oxide, ammonia, and the hydrocarbons: butane, propane, ethane, ethylene, and methane. In a recent study, Smith *et al.*[109] used supercritical ammonia for supercritical fluid extraction and fractionation of marine diesel fuel and storage sediments, but concern has been expressed about corrosion. Ethers, particularly isopropyl ether and diethyl ether have also been used as mobile phases. Halogenated methanes have been used in a number of studies. These include freon 23, freon 11, CCl_2F_2, and $HCClF_2$.[6] In one interesting paper, French and Novotny described the use of supercritical xenon as a unique mobile phase in conjunction with FTIR detection.[110] To alter selectivities, organic modifiers can be added to the mobile phase. Alcohols have been the most widely used mobile phase organic modifier used in SFC, and among them, methanol has been predominant. Other modifiers include dioxane, acetonitrile, tetrahydrofuran, di-isopropyl ether and diethyl ether. Most commercial capillary column SFC systems use carbon dioxide as the mobile phase.

Detectors

Conventional detectors from both GLC and HPLC have been successfully adapted to SFC (Table 3). The ultraviolet (UV) spectrophotometric detector has been the most widely used in SFC followed by the flame ionization detector (FID), although this is likely to change with the introduction of commercial capillary SFC instruments with built-in FID. There has been considerable interest in the use of SFC with mass spectrometric or Fourier transform mass spectrometric detectors (FTMS), and infrared or Fourier transform infrared spectrophotometric detectors (FTIR). Other detectors used in SFC include the fluorescence spectrophotometric detector, flame photometric detector (FPD), thermionic detector (TID), and photoionization detector (PID).

Table 3 *Detectors commonly used in SFC*

Detector	Number of papers*
UV	88
FID	64
MS/FTMS	28
IR/FTIR	17
Fluorescence detector	9
Flame photometric detector	3
Thermionic detector	2

* Based on a survey, July, 1987.

The flame ionization detector has been widely used in conjunction with such mobile phases as CO_2 or N_2O, which yield negligible background signals. This detector, however, is likely to be incompatible with the addition of greater than ten percent of organic modifier in the mobile phase. Fluorescence spectroscopy is an attractive method for the detection of polynuclear aromatic compounds. The high selectivity and sensitivity of the thermionic detector for phosphorous and nitrogen-containing substances have made it very attractive in GLC for applications in biochemical and clinical chemistry, such as in the analyses of amino acids, drugs, and pharmaceuticals. However, it appears that, apart from two recent reports[111,112] there has not been much work on SFC using thermionic detection. Despite having the potential of being highly sensitive, few other established selective chromatographic detection methods, such as the electron capture detector or electrochemical detector, have been explored for use in SFC.

Different types of detector can be used simultaneously either in parallel or in tandem in order to obtain better selectivity of detection and/or structural identification of the analytes. Some of the detector combinations used in SFC reported in the literature include UV-FID, UV-MS, UV-FTIR, FID-MS, FID-FPD, and FID-TID. Several workers have explored alternative detection methods for SFC; Jinno *et al.*[113,114] utilized a multichannel UV detector, Asche[115] used a refractive index detector, Fjeldsted *et al.*[116] used a scanning fluorescence detector, Gluckman, Shelly, and Novotny[117] used a fluorescence photodiode array, and Cashaw *et al.*[118] used a microadsorption detector.

Commercial *vs.* Home-made SFC Systems

Dedicated SFC instruments are now commercially available from a number of manufacturers including Lee Scientific, Brownlee Labs, Computer Chemical Systems (CCS), Jasco, LDC, and Carlo Erba. The design and operation of one such system, the Lee Scientific Model 501 capillary supercritical fluid chromatograph, has been discussed by Later *et al.*[119] So far, however, it appears that work on SFC using commercial SFC instruments, as suggested by the number of published papers, is rather limited. This can probably be explained by a number of factors: (*a*) primarily, these commercial SFC instruments have only recently been introduced as established equipment, attractive to pharmaceutical and petroleum industries; (*b*) users have the alternative of constructing their own SFC systems by making (often non-permanent) modifications to conventional GC or HPLC instruments; (*c*) SFC has not yet been widely accepted as a routine analytical technique in laboratories, hence maintaining the low demand for commercial SFC systems; (*d*) considerable technological problems in SFC have yet to be solved; (*e*) commercial SFC systems including the associated computers are relatively expensive; and (*f*) many users in new technology areas are reluctant to release commercial details of the applications (particularly on commercial instruments).

Numerous papers have described modifications made to existing GC or HPLC instruments for use in SFC. Grob[120] described an attempt to extend the use of capillary GC to capillary SFC without drastically modifying the equipment. However, he found that the feasible working conditions were rather limited. In 1982, Gere, Board, and McManigill[121] described the modification of a Hewlett Packard 1084B liquid chromatograph (with eluent gradient at a constant flow-rate capability) to permit operation with liquid carbon dioxide at the pump stage while maintaining supercritical conditions at the separation stage. A similar SFC system was made commercially available for a limited period, and subsequently, several workers[122-124] performed the same modifications using this HPLC instrument. In most cases, the SFC systems have been assembled by modification of different commercial components, in particular the mobile phase delivery system (see next section).

Mobile Phase Delivery Systems in SFC

Liquefied gases such as carbon dioxide and nitrous oxide are normally available from the cylinder and they are often available in various grades. These gases in the liquid state can be drawn out of the cylinder either by using an aductor (dip) tube, or by turning the cylinder upside down.[111,125] The mobile phase is then transferred to the rest of the SFC system using a suitable tubing such as 1/16 inch stainless steel tubing. A fine filter (less than 5 μm) placed in line before the pump will free the liquid from any coarse particles from the cylinder which may contaminate the pump and the system. In the case where CO_2 is used as mobile phase, it may be necessary to add, prior to the injector, a purifier such as an adsorbent trap (usually made of activated carbon or alumina) to improve the CO_2 quality. This extra column can also act as an efficient pulse damper.[126]

High pressures, greater than that supplied by the gas cylinder, can be achieved by pressurising the cylinder by careful heating. Although there are obviously some limitations to this technique, pressures of up to 150 bar can be conveniently obtained by heating a CO_2 cylinder to about 55–60 °C with necessary pressure control obtained by means of a pressure valve.[9,120] However, the vast majority of work in SFC, has used a high pressure pump to achieve higher pressures. Reciprocating pumps have generally been used in conjunction with packed-column SFC. These pumps, besides having the advantage of being able to deliver unlimited volumes with a continuous flow of mobile phase, many also incorporate eluent gradient capability.[126] Problems with pulsating flow associated with these pumps can be reduced to some extent by the use of pulse dampers.

The density of supercritical fluids varies with pressure, with the variation greatest in the region near the critical point. Because the pump works against a restrictor in order to maintain the desired pressure in the column, pressure control is generally more important than flow control. Most previous workers have used conventional HPLC pumps after they were modified for pressure control.[121,125] Pressure gradients in SFC have been accomplished in different ways, depending on the method of supercritical fluid delivery.

For an SFC system that uses a reciprocating pump, a pressure programmer can be employed. The pressure programmer described by Jentoft and Gouw[22] consisted of a cylinder of high pressure nitrogen, variable motors, and pressure regulators which essentially controlled the inlet pressure to the pump. Van Lanten and Rothman[127] described a pump pressure control design which involved electronics controlled by a microcomputer, applicable to almost any commercial LC pump.

Because of the low flow rates requirement (several µl min^{-1}), syringe pumps have been used for most capillary SFC. These pumps have the principal advantage of delivering pulseless flow, thus offering better baseline stability and a lower detection limit than that of reciprocating pumps. Furthermore, they are known to be generally reliable and easy to operate.[126]

It is usually necessary to cool the pump head when working with eluents that are gaseous at ambient temperatures. Cooling reduces the tendency of the liquid eluent to undergo gasification or cavitation in the pump head. Gere *et al.*[121] noted that cooling the pump heads to 0 °C or below greatly enhanced the pumping efficiency. Several techniques for cooling the pump head have been described or suggested in the literature. An interesting approach suggested by Bruno is cooling the pump components by using a simple vortex refrigeration method.[128] He used a vortex tube which can be operated by a source of compressed air and temperatures as low as -15 °C are easily reached. This technique, however, has not gained acceptance in SFC, probably because of its limited cooling applicability to small-scale components and its rather noisy operation.

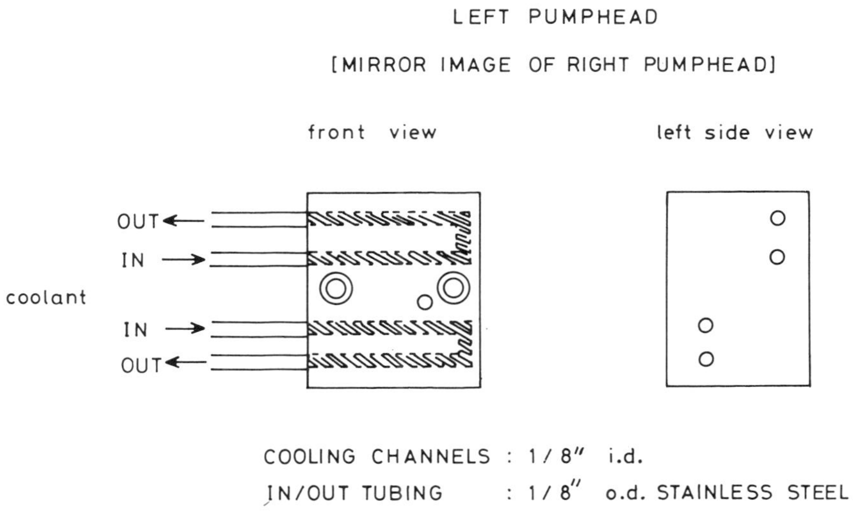

LEFT PUMPHEAD

[MIRROR IMAGE OF RIGHT PUMPHEAD]

COOLING CHANNELS : 1 / 8″ i.d.

IN/OUT TUBING : 1 / 8″ o.d. STAINLESS STEEL

Figure 2 *Schematic diagram of the pump head cooling system for a Waters Associates Model 6000 A pump with square pump heads for the pumping of liquid* CO_2 *(Reproduced by permission from Anal. Chem., 1984,* **56**, *2681, Copyright 1984, American Chemical Society.)*[125]

Greibrokk *et al.*[125] described the modification of a Waters Associates Model 6000A pump with square pump heads. Cooling channels were drilled in the solid outer parts of both pump heads (Figure 2). Two additional check valves (inlet and outlet) were placed in a home-made block of stainless steel containing cooling channels (Figure 3). Cooling of the pump heads and the check valves was accomplished by running cold methanol ($-8\,^{\circ}$C) through the cooling channels.

COOLING BLOCK FOR TWO SSI CHECK-VALVES

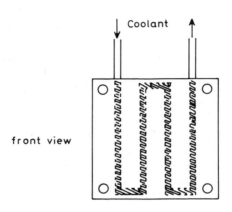

front view

MATERIAL   : STAINLESS STEEL
COOLING CHANNELS : 1/8″ i.d.
IN/OUT TUBING   : 1/8″ o.d.

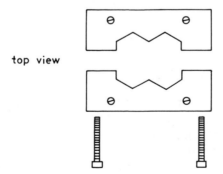

top view

Figure 3 *Schematic diagram of the check valve cooling system for a Waters Associates Model 6000 A pump for the pumping of liquid* CO_2
(Reproduced by permission from *Anal. Chem.*, 1984, **56**, 2681, Copyright 1984, American Chemical Society.)[125]

In their SFC investigations Gere, Board, and McManigill[121] cooled the two liquid pumping heads of a HP 1084B liquid chromatograph by placing them in close thermal contact with clamp-on heat exchangers, through which a water-methanol mixture from an external circulating chiller was circulated. Simpson, Gant, and Brown[129] accomplished cooling of the pump heads of Perkin-Elmer Series 10 pumps by circulating ice water through a heat exchanger plate in thermal contact with the pump heads. The heat exchanger they used was a hollow copper plate with coolant inlet and outlet ports and it was secured to the pump head by the head retaining screws which were passed through appropriately placed holes in solid areas of the exchanger. Greibrokk *et al.*[111] described in detail the construction of a clamp-on cooling unit for Waters Model 590 pump. The cooling unit was made from aluminium and consisted of three pieces with the centre piece precisely cut to fit accurately between the two pump heads (Figure 4).

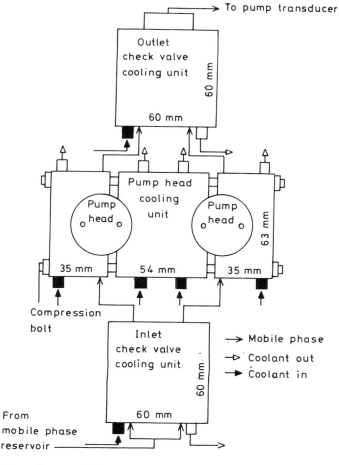

Figure 4 *Flow diagram of the pump head and check valve cooling system for a Waters Model 590 pump for the pumping of liquid* CO_2
(Reproduced from *J. Chromatogr.*, 1986, **371**, 145, by permission of the authors and Elsevier Scientific Publishing Co.)[111]

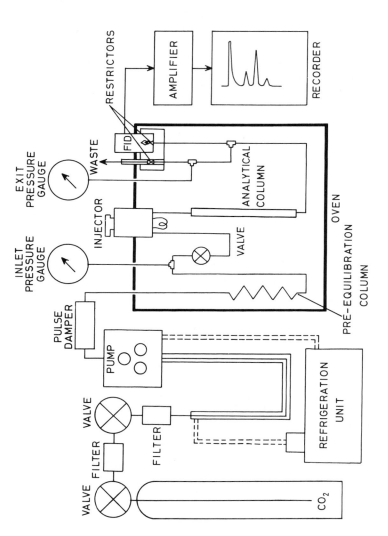

Figure 5 *Schematic diagram of a packed-column supercritical fluid chromatographic system with flame ionization detector* (Adapted from ref. 132, *Anal. Proc.*, 1987, **24**, 304)[132]

Cooling channels were drilled through each separate part and cooling was effected by running cold methanol through them. A similar technique was also reported by several workers. Rawdon[130] used a 'cooling cap' through which cold water-glycol solution was circulated. Mourier, Eliot, Caude, Rosset, and Tambute[131] used a "clamp-on" heat exchanger through which cold ethanol was circulated.

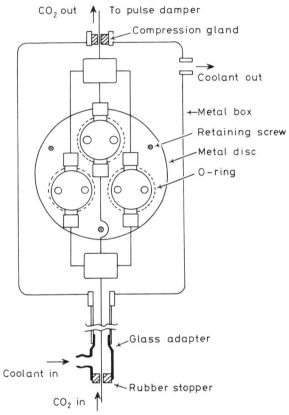

Figure 6 *Schematic diagram of the pump head, check valves, and solvent mixtures cooling unit for the Jasco Familic 300S HPLC pump for the pumping of liquid* CO_2

In our laboratory, a packed-column supercritical fluid chromatograph has been constructed from an HPLC pump and a Pye 104 gas chromatograph fitted with an effluent splitter and a flame ionization detector.[132,133] Figure 5 shows a schematic diagram of the SFC system. The three circular-shaped pump heads, check valves, and solvent mixers of a Jasco Familic 300S HPLC pump were modified to allow supercritical fluid chromatographic operation with CO_2 as mobile phase.[133] Cooling of the pump heads, check valves, and mixers was accomplished by enclosing the components in a box through which cold (*ca.* -20 to $-5\,^\circ$C) ethylene glycol-water mixture was run (Figure 6). The cooling box was made from an ordinary diecast waterproof electrical box which was large enough to accommodate the pump heads, check valves, and solvent mixers. The back part of the box was precision-cut to fit the pump heads. Water-tight seals between the pump heads and the box were provided by O-rings. Inlet and outlet ports on the box allowed the circulation of the coolant. Because there was a direct contact between the coolant and the pump heads, cooling of the latter was very effective. The cooling efficiency was further improved by cooling a major portion of the CO_2 inlet tubing prior to the pump. This was made possible by enclosing the length of tubing in an ordinary rubber tube (3/8 inch i.d.), one end of which led to the coolant pump, and the other attached directly to the inlet port of the cooling box.

In the case of syringe pumps, cooling is not always essential. A number of workers[134-136] reported the use of Varian Model 8500 high pressure syringe pumps without any cooling. However, cooling can be employed to enhance the filling of the pump during a refill cycle. Doehl et al.[137] in their SFC studies used an ISCO µLC-500 syringe pump which was equipped with a cooling coil through which was circulated methanol at $-15\,°C$ during the pump refill procedure. This syringe pump has become the basis of the delivery systems in many of the present commercial SFC instruments.

In the process of cooling the pump heads, it is important to provide good insulation to the entire cooling system to prevent excessive air moisture condensation and loss of cooling efficiency. Cooling units can be insulated using cut-to-fit polystyrene foam, and expanded polystyrene hoses can provide insulation to the cooling system tubing.

The cooling of the pump head and components may create some adverse effects on the pump performance. Some makes of pump seals, when operated under cold conditions tend to develop leaks of the mobile phase (CO_2) and fail to perform satisfactorily. We observed this effect when using a Kontron 1070 LC pump, and Greibrokk et al.[111,137] had similar problems with a Milton Roy microMetric metering pump and an ISCO µLC-500 syringe pump. Thus for the latter, cooling was turned on only during the refilling procedure since leaks were observed through cold piston seals.

Organic modifiers can be introduced into the eluent system in a number of ways. Several workers[122,124,138,139] used cylinders with CO_2 doped with methanol, 2-propanol, or other modifiers (e.g. 10% w/w) which are commercially available. These cylinders can be used as received. However, to permit more accurate addition of small amounts of modifier to the CO_2 mobile phase, further dilution is possible by plumbing the doped CO_2 into one pump (say pump 'B') of a dual headed pump.[124] Alternatively, various mixtures of modifier in CO_2 can be made by mixing them in a lecture bottle or a syringe pump, according to the techniques suggested by Yonker et al.[139,140] Another method reported in the literature for adding modifier to CO_2 was by direct on-line mixing with the use of another pump. Greibrokk et al. accomplished this with the use of a Waters Model 590 microflow pump and a tubing T-piece[141] or a home-made high pressure low-volume mixer.[125] Microflow pumps such as Waters Model 590 and Gilson Model

302 are able to pump at low flow-rates (less than $5\,\mu l\,min^{-1}$). A T-piece mixing system may be adequate for most analytical purposes as suggested by the low retention time variations of less than 2%.[141] Novotny and David[142] used a dual syringe Brownlee Labs micropump to add 5% methanol into butane and the mixing was accomplished with a Brownlee mixing cartridge (mixer-52) inserted in-line prior to the injection valve.

Future Developments in Instrumentation

There has been considerable development in the SFC industry and it is expected to maintain its rapid growth in the coming years. The likely changes which we can expect would be the establishment of commercial packed-column SFC instruments, and the advancement of packed-capillary column technology for use in SFC. The supercritical fluid chromatographic application of fused-silica capillary packed columns (*e.g.* 0.2 mm i.d. × 50 cm packed with Finesil C18-10 of Jasco) has been under rigorous investigation by Hirata and his co-workers.[143,144] Two interesting areas which have shown great potential for further development are the direct coupling of supercritical fluid extraction to SFC, and the extension of SFC to semi-preparative or preparative scale supercritical fluid chromatographic separations. Instruments for the former have been introduced into the market (*e.g.* Jasco Model Super-100) but the numbers appear to be rather limited.

4 Applications and Challenges in SFC

After about five years of rapid growth, SFC continues to attract consider-able attention. A wide variety of applications have been investigated in analytical, as well as in industrial sectors. SFC has been utilized for practical applications ranging from the analysis of oil residues[145] and polynuclear aromatic hydrocarbons in exhausts[24] to pharmaceuticals[135] and the more exotic separation of beeswax.[146]

Hydrocarbons and petroleum derivatives have been the subject of many investigations. The supercritical fluid chromatographic separation of ethers, ketones, aldehydes, and more polar compounds including organic acids and free fatty acids, as well as lipids and glycerides have also been investigated. The number of reported papers describing work on SFC with polynuclear aromatic compounds and/or polystyrene oligomers as analytes for test solutes is overwhelming (no less than 54 and 21 papers respectively; Table 4). Polystyrene contains a series of oligomers having a wide range of molecular weights; therefore, it has been used more or less as a standard to estimate the efficiency of the chromatographic system.[147]

Looking at the tremendous development, we can expect SFC to have a bright future although it has to strive to establish a firm ground between GLC and HPLC[107] (see also Chapters 4 and 5). Capillary SFC with spectrophotometric detection has been used successfully in the analysis of polar compounds with advantages over HPLC methods, and it has also been used in the separation of thermally labile compounds not amenable to GLC. However, there has been some criticism that SFC has not yet proved its merit as a chromatographic tool because a large majority of the reported applications of SFC can be carried out either on GLC or HPLC, with equal or better resolution and greater ease of operation.

Table 4 *Examples of the applications of SFC (1962–mid-1987)*

Substance determined or Material analysed	Number of papers
Polystyrene oligomers	21
Polynuclear aromatic hydrocarbons	54
Hydrocarbons	25
Coal and petroleum derivatives, refinery residues, diesel, kerosine, gasoline, coal liquefactions	12
Carboxylic acids	2
Alcohols	7
Aldehydes, ketones	4
Fatty acids, fatty acid esters, lipids, glycerides	9
Carbohydrate and sucrose esters	3
Vitamins	2
Pesticides, carbamates, herbicides	8
Drugs: caffeine, codeine, methocarbamol, oxyphenbutazone, phenylbutazone, theophylline, trichothecene mycotoxins	7
Erythromycin	1
Prostaglandins	1
Metal chelates	4
GLC stationary phases	2
Phosphine oxide enantiomers	2
S- or P-containing compounds in cigarette smoke, pesticides	2
Halogenated compounds	2
Foods: vegetable and butter oils, cheese, coffee, tobacco, chamomile	3
Liquid crystal	1
Waxes	2
Insect sex hormone	1
Terpenes	1
Phosphine oxide enantiomers	2
Amino acid enantiomers	1

Today, SFC faces a strong competition from the high-temperature GLC, which was introduced only recently after technological advances in fused silica capillary columns. By replacing the polyimide outer coat of blank fused silica capillary columns with aluminium (forming an aluminium-clad capillary column) the maximum operating temperature of the columns has been increased from *ca.* 370 °C to 600 °C.[148] A thin film of bonded and cross-linked non-polar stationary phase on the new column has been successfully prepared using a semi-viscous, high molecular-weight silanol-terminated poly(dimethyl siloxane). This column can be operated iso-thermally to about 415 °C and can be temperature-programmed for short periods to about 440 °C. Many of the separations of high molecular-weight compounds carried out by SFC have been demonstrated to be feasible now using GLC, see Chapter 5.

The role of SFC lies in its applications. Chester[149] examined the role of SFC in analytical problem solving and defined this role of SFC with respect to the capabilities and limitations of GLC and HPLC. Many authors have described SFC as a link between GLC and HPLC. However, as Schoen-makers has highlighted,[150] there has often been an unfortunate miscon-ception made in papers on SFC, 'to assign the best of everything to a supercritical fluid'. This mistaken claim has led to the portrayal of SFC as having the 'magical' solution for many difficult separations. But, as commented by Hawkes in his recent article,[151] analysts should obviously choose the most suitable technique for any given problem.

The current technological challenges in capillary SFC have been discussed by Lee and Markides.[50] They have also expressed the future need for developments in SFC instrumentation, particularly in sample introduction methods, column and restrictor technology, and utilization of polar mobile phases. Recently, there have been calls for rigorous exploration of new research areas which possibly will lead to unique applications of SFC.[107,151] The prospects of SFC as a competitive and viable separation technique, and some of the possible research goals in SFC have been clearly outlined by Schoenmakers.[107] Accordingly, to assure a continuous growth of SFC, new developments in instrumentation and applications of this technique must be sought and, in the application aspect at least, it seems that SFC has to confront a challenging future.

References

1. L. Blomberg and T. Wannman, *J. Chromatogr.*, 1979, **168**, 81.
2. R. P. W. Scott, P. Kucera, and M. Munroe, *J. Chromatogr.*, 1979, **186**, 475.
3. Y. Hirata and M. Novotny, *J. Chromatogr.*, 1979, **186**, 521.
4. A. J. Martin and R. L. M. Synge, *Biochem. J.*, 1941, **35**, 1358.
5. J. B. Hannay and J. Hogarth, *Proc. Roy. Soc., (London)*, 1879, **29**, 324, *ibid.*, 1880, **30**, 178.
6. E. Klesper, A. H. Corwin, and D. A. Turner, *J. Org. Chem.*, 1962, **27**, 700.
7. N. M. Karayannis and A. H. Corwin, *J. Chromatogr.*, 1970, **47**, 247.
8. N. M. Karayannis and A. H. Corwin, *J. Chromatogr. Sci.*, 1970, **8**, 251.

9. S. T. Sie, W. Van Beersum, and G. W. A. Rijnders, *Sep. Sci.*, 1966, **1**, 459.

10. S. T. Sie and G. W. A. Rijnders, *Sep. Sci.*, 1967, **2**, 699.

11. S. T. Sie and G. W. A. Rijnders, *Sep. Sci.*, 1967, **2**, 755.

12. S. T. Sie and G. W. A. Rijnders, *Anal. Chim. Acta*, 1967, **38**, 31.

13. J. C. Giddings, W. A. Manwaring, and M. N. Myers, *Science*, 1966, **154**, 146.

14. J. C. Giddings, *Sep. Sci.*, 1966, **1**, 73.

15. M. N. Myers and J. C. Giddings, *Sep. Sci.*, 1966, **1**, 761.

16. L. B. Bowman, Jr., M. N. Myers, and J. C. Giddings, *Sep. Sci. Technol.*, 1982, **17**, 271.

17. L. McLaren, M. N. Myers, and J. C. Giddings, *Science*, 1968, **159**, 197.

18. J. C. Giddings, M. N. Myers, and J. W. King, *J. Chromatogr. Sci.*, 1969, **7**, 276.

19. J. J. Kirkland, *J. Chromatogr. Sci.*, 1969, **7**, 7.

20. T. Altares, Jr., *J. Polym. Sci., Part B*, 1970, **8**, 761.

21. R. E. Jentoft and T. H. Gouw, *J. Chromatogr. Sci.*, 1970, **8**, 138.

22. R. E. Jentoft and T. H. Gouw, *Anal. Chem.*, 1972, **44**, 681.

23. R. E. Jentoft and T. H. Gouw, *J. Polymer Sci., Part B*, 1969, **7**, 811.

24. R. E. Jentoft and T. H. Gouw, *Anal. Chem.*, 1976, **48**, 2195.

25. J. A. Nieman and L. B. Rogers, *Sep. Sci. Technol.*, 1975, **10**, 517.

26. M. Novotny, S. R. Springston, P. A. Peaden, J. C. Fjeldsted, and M. L. Lee, *Anal. Chem.*, 1981, **53**, 407A.

27. O. Vitzthum, P. Huber, and M. Barthels, *Ger. Offen.*, Pat. No. 2159339, (*Chem. Abstr.*, 1973, **79**, 048966).

28. W. Hartmann and E. Klesper, *Ger. Offen.*, Pat. No. 2729462, (*Chem. Abstr.*, 1979, **90**, 154127).

29. M. Perrut, *Fr. Demande*, Pat. No. 2527934, (*Chem. Abstr.*, 1984, **100**, 105823)

30. Yokogawa-Hewlett Packard Ltd., *Jpn. Kokai Tokkyo Koho*, Pat. No. 83 17364, (*Chem. Abstr.*, 1983, **99**, 115202).

31. Morinaga Confectionary Co. Ltd., *Jpn. Kokai Tokkyo Koho*, Pat. No. 85 08747, (*Chem. Abstr.*, 1985, **103**, 016219).

32. D. W. Vidrine and D. R. Allhands, U.S. Pat. No. 4588893, (*Chem. Abstr.*, 1986, **105**, 145338).

33. M. Novotny, M. L. Lee, P. A. Peaden, J. C. Fjeldsted, and S. R. Springston, U.S. Pat. No. 4479380, (*Chem. Abstr.*, 1985, **102**, 72096)

34. Anonymous, *Anal. Chem.*, 1987, **59**, 619A.

35. J. C. Giddings, M. N. Myers, L. McLaren, and R. A. Keller, *Science*, 1968, **162**, 67.

36. T. H. Gouw and R. E. Jentoft, *J. Chromatogr.*, 1972, **68**, 303.

37. G. Schomburg, *Ber. Bunsenges. Phys. Chem.*, 1973, **77**, 219.

38. Jun-Shung Huang, *Hua Hsueh Tung Pao*, 1974, (3), 50. (*Chem. Abstr.*, 1974, **81**, 130601).

39. A. Przyjazny and R. Ataszewski, *Chem. Anal. (Warsaw)*, 1974, **19**, 3. (*Chem. Abstr.*, 1974, **81**, 45021).

40. T. H. Gouw and R. E. Jentoft, *in* 'Advances in Chromatography'. J. C. Giddings, E. Grushka, R. A. Keller, and J. Cazes (*eds.*), Vol. 13, Marcel Dekker, New York, 1975, pp. 1–40.
41. E. Klesper, *Angew. Chem., Int. Ed. Engl.*, 1978, **71**, 738.
42. T. H. Gouw, R. E. Jentoft, and E. J. Gallegos, *High-pressure Sci. Technol., AIRAPT Conf., 6th*, 1979, **1**, 583.
43. T. H. Gouw and R. E. Jentoft, *Chromatogr. Sci.*, 1979, **11**, 313.
44. M. V. Novotny, *Proc. Int. Symp. Capillary Chromatogr., 4th.*, 1981, 177.
45. J. C. Fjeldsted and M. L. Lee, *Anal. Chem.*, 1984, **56**, 619A.
46. M. Novotny, *Analyst*, 1984, **109**, 199.
47. M. Novotny, *in* 'Microcolumn Separations', M. V. Novotny and D. Ishii (*eds.*), *J. Chromatogr. Libr.*, Vol. **30**, Elsevier, Amsterdam, 1985, pp. 105–120.
48. M. Caude and R. Rosset, *Analusis*, 1986, **14**, 310.
49. D. W. Later, B. E. Richter, and M. R. Andersen, *LC-GC*, 1986, **4**, 992.
50. M. L. Lee and K. E. Markides, *J. High Resolut. Chromatogr., Chromatogr. Commun.*, 1986, **9**, 652.
51. K. E. Markides and M. L. Lee, *in* 'Advances in capillary chromatography', J. G. Nikelly (*ed.*), Huethig, Heidelberg, 1986, pp. 19–34.
52. R. D. Smith, B. W. Wright, and H. R. Udseth, *ACS Symp. Ser.*, 1986, **297**, 260.
53. B. W. Wright, H. T. Kalinoski, H. R. Udseth, and R. D. Smith, *J. High Resolut. Chromatogr., Chromatogr. Commun.*, 1986, **9**, 145.
54. D. Richter and D. W. Later, *Lab. Sci. Technol.*, 1986, **(11)**, 17.
55. H. T. Kalinoski, H. R. Udseth, E. K. Chess, and R. D. Smith, *J. Chromatogr.*, 1987, **393**, 3.
56. J. W. Jordan, R. J. Skelton, and L. T. Taylor, *ACS Symp. Ser.*, 1987, **329**, 189.
57. K. Nurmela, *Kem. Kemi*, 1983, **10**, 795.
58. L. G. Randall, *Sep. Sci. Technol.*, 1982, **17**, 1.
59. K. Nagama, *Kagaku to Kogyo (Tokyo)*, 1984, **37**, 318. (*Chem. Abstr.*, 1986, **104**, 179138).
60. D. Ishii and T. Takeuchi, *Kemikaru Enjiniyaringu*, 1985, **30**, 453. (*Chem. Abstr.*, 1985, **103**, 115335).
61. M. Saito and T. Hondo, *Shokuhin Kikai Sochi*, 1985, **22**, 45. (*Chem. Abstr.*, 1986, **104**, 33084).
62. M. Saito and T. Hondo, *Yukagaku*, 1986, **35**, 273. (*Chem. Abstr.*, 1986, **104**, 226876).
63. G. M. Schneider, *Angew. Chem., Int. Ed. Engl.*, 1978, **17**, 716.
64. U. Van Wassen, I. Swaid, and G. M. Schneider, *Angew. Chem., Int. Ed. Engl.*, 1980, **19**, 575.
65. P. A. Peaden and M. L. Lee, *J. Liq. Chromatogr.* 1982, **5** (Suppl. 2), 179.
66. E. Klesper and D. Leyendecker, *Int. Lab.*, 1986, **16** (9), 18.

67. C. R. Yonker, B. W. Wright, H. R. Udseth, and R. D. Smith, *Ber. Bunsenges. Phys. Chem.*, 1984, **88**, 908.
68. G. M. Schneider, *Ber. Bunsenges. Phys. Chem.*, 1984, **88**, 481.
69. G. M. Schneider, *Thermochim. Acta*, 1985, **88**, 17.
70. F. P. Schmitz and E. Klesper, *J. Chromatogr.*, 1987, **388**, 3.
71. E. Klesper and F. P. Schmitz, *J. Chromatogr.*, 1987, **402**, 1.
72. M. Novotny, *J. High Resolut. Chromatogr., Chromatogr. Commun.*, 1986, **9**, 137.
73. J. Cousin and P. J. Arpino, *Analusis*, 1986, **14**, 215.
74. T. R. Covey, E. D. Lee, A. P. Bruins, and J. D. Henion, *Anal. Chem.*, 1986, **58**, 1451A.
75. D. E. Games, A. J. Berry, I. C. Mylchreest, J. R. Perkins, and S. Pleasance, *Lab. Pract.*, 1987, **36**, 45.
76. D. R. Gere, *Science*, 1983, **222**, 253.
77. R. D. Smith, B. W. Wright, and H. R. Udseth, *in* 'Advances in Capillary Chromatography', J. G. Nikelly (*ed.*), Huethig, Heidelberg, 1986, pp. 56–94.
78. P. R. Griffiths, S. L. Pentony, Jr., A. Giorgetti, and K. H. Shafer, *Anal. Chem.*, 1986, **58**, 1349A.
79. D. Weiboldt and G. Adams, *Int. Labmate Guide*, 1987, p. 11.
80. K. Jinno, *Chromatographia*, 1987, **23**, 55.
81. K. D. Bartle, *Lab. Pract.*, 1985, **34**, 59.
82. R. E. Clement, F. I. Onuska, G. A. Eiceman, and H. H. Hill, Jr., *Anal. Chem.*, 1986, **58**, 321R.
83. H. Engelhardt, *Fresenius' Z. Anal. Chem.*, 1987, **317**, 27.
84. K. Fujita, *Yukagaku*, 1983, **32**, 551. (*Chem. Abstr.*, 1983, **99**, 224396).
85. K. Fujita, *Fragrance J.*, 1984, **12**, 50. (*Chem. Abstr.*, 1985, **102**, 155250).
86. K. Fujita, *Bunseki*, 1985, (12), 847. (*Chem. Abstr.*, 1986, **104**, 179417).
87. J. L. Hensley, E. Leaseburge, and H. M. McNair, *Adv. Instrum.*, 1985, **40**, 287.
88. W. F. Hsieh, *Hua Hsueh*, 1985, **43**, 39. (*Chem. Abstr.*, 1986, **104**, 161028).
89. W. P. Jackson, B. E. Richter, J. C. Fjeldsted, R. C. Kong, and M. L. Lee, *ACS Symp. Ser.*, 1984, **250**, 121.
90. F. M. Lancas, *Rev. Quim. Ind., (Rio de Janeiro)*, 1986, **55**, 9. (*Chem. Abstr.*, 1986, **105**, 71800).
91. M. L. Lee and K. E. Markides, *Science*, 1987, **235**, 1342.
92. P. Mourier, M. Caude, and M. Rosset, *Analusis*, 1984, **12**, 229.
93. P. Mourier, *Analusis*, 1987, **15**, 200.
94. R. D. Smith, B. W. Wright, and H. R. Udseth, *Anal. Chem. Symp. Ser.*, 1984, **19**, 375.
95. T. Takeuchi and D. Ishii, *Kagaku no Ryoiki Zokan*, 1983, (138), 59. (*Chem. Abstr.*, 1983, **99**, 128686).
96. C. M. White and R. K. Houck, *J. High Resolut. Chromatogr., Chromatogr. Commun.*, 1986, **9**, 4.
97. W. Ecknig and H. J. Polster, *Mitteilungsbl. Chem. Ges. D.D.R.*, 1986, **33**, 225. (*Chem. Abstr.*, 1987, **106**, 95279).

98. Y. Hirata and S. Tsuge, *Kagaku (Kyoto)*, 1986, **41**, 130. (*Chem. Abstr.*, 1987, **106**, 130967).

99. E. Klesper and F. P. Schmitz, *Chim. Oggi*, 1986, (11), 11. (*Chem. Abstr.*, 1987, **106**, 43161).

100. Z. Wang, *Fenxi Huaxue*, 1984, **12**, 1025–31, 996. (*Chem. Abstr.*, 1985, **102**, 55337).

101. M. Novotny, *in* 'The Science of Chromatography', F. Bruner (*ed.*), *J. Chromatogr. Libr.*, **32**, Elsevier, Amsterdam, 1985, pp. 305–332.

102. M. Novotny, *J. Pharm. Biomed. Anal.*, 1984, **2**, 207.

103. G. B. Levy, *Am. Lab.*, 1986, **18**, 62.

104. M. Novotny, *J. High Resolut. Chromatogr., Chromatogr. Commun.*, 1987, **10**, 248.

105. T. Greibrokk, B. E. Berg, A. L. Blilie, J. Doehl, A. Farbrot, and E. Lundanes, *J. Chromatogr.*, 1987, **394**, 429.

106. H. E. Schwartz, *LC & GC*, 1987, **5**, 14.

107. P. J. Schoenmakers and C. C. J. G. Verhoeven, *Trends Anal. Chem.*, 1987, **6**, 1.

108. H. E. Schwartz, P. J. Barthel, S. E. Moring, and H. H. Lauer, *LC-GC*, 1987, **5**, 490.

109. R. D. Smith, H. R. Udseth, and R. N. Hazlett, *Fuel*, 1985, **64**, 810.

110. S. B. French and M. Novotny, *Anal. Chem.*, 1986, **58**, 164.

111. T. Greibrokk, J. Doehl, A. Farbrot, and B. Iversen, *J. Chromatogr.*, 1986, **371**, 145.

112. W. R. West and M. L. Lee, *J. High Resolut. Chromatogr., Chromatogr. Commun.*, 1986, **9**, 191.

113. K. Jinno, T. Hondo, and M. Saito, *Chromatographia*, 1985, **20**, 351.

114. K. Jinno, T. Hoshino, T. Hondo, M. Saito, and M. Senda, *Anal. Chem.*, 1986, **58**, 2696.

115. W. Asche, *Chromatographia*, 1987, **11**, 411.

116. J. C. Fjeldsted, B. E. Richter, W. P. Jackson, and M. L. Lee, *J. Chromatogr.*, 1983, **279**, 423.

117. J. C. Gluckman, D. C. Shelly, and M. V. Novotny, *Anal. Chem.*, 1985, **57**, 1546.

118. J. L. Cashaw, R. Segura, and A. Zlatkis, *J. Chromatogr. Sci.*, 1970, **8**, 363.

119. D. W. Later, B. E. Richter, W. D. Felix, and M. R. Andersen, *Am. Lab.*, 1986, **18**, 108.

120. K. Grob, *J. High Resolut. Chromatogr., Chromatogr. Commun.*, 1983, **6**, 178.

121. D. R. Gere, R. Board, and D. McManigill, *Anal. Chem.*, 1982, **54**, 730.

122. J. M. Levy and W. M. Ritchey, *J. High Resolut. Chromatogr., Chromatogr. Commun.*, 1985, **8**, 503.

123. D. Leyendecker, D. Leyendecker, F. P. Schmitz, and E. Klesper, *J. High Resolut. Chromatogr., Chromatogr. Commun.*, 1986, **9**, 525.

124. J. B. Crowther and J. D. Henion, *Anal. Chem.*, 1985, **57**, 2711.

125. T. Greibrokk, A. L. Blilie, E. J. Johansen, and E. Lundanes, *Anal. Chem.*, 1984, **56**, 2681.

126. J. F. K. Huber, (*ed.*), 'Instrumentation for High Performance Liquid Chromatography', *J. Chromatogr. Libr.*, **13**. Elsevier, Amsterdam, 1978.
127. F. J. Van Lanten and L. D. Rothman, *Anal. Chem.*, 1976, **48**, 1430.
128. T. J. Bruno, *LC Magazine*, 1986, **4**, 134.
129. R. C. Simpson, J. R. Gant, and P. R. Brown, *J. Chromatogr.*, 1986, **371**, 109.
130. M. G. Rawdon, *Anal. Chem.*, 1984, **56**, 831.
131. P. A. Mourier, E. Eliot, M. H. Caude, R. H. Rosset, and A. G. Tambute, *Anal. Chem.*, 1985, **57**, 2819.
132. M. M. Sanagi and R. M. Smith, *Anal. Proc.*, 1987, **24**, 304.
133. R. M. Smith and M. M. Sanagi, *J. Pharm. Biomed. Anal.*, in press.
134. P. A. Peaden, J. C. Fjeldsted, M. L. Lee, S. R. Springston, and M. Novotny, *Anal. Chem.*, 1982, **54**, 1090.
135. B. E. Richter, *J. High Resolut. Chromatogr., Chromatogr. Commun.*, 1985, **8**, 297.
136. T. L. Chester, D. P. Innis, and G. D. Owens, *Anal. Chem.*, 1985, **57**, 2243.
137. J. Doehl, A. Farbrot, T. Greibrokk, and B. Iversen, *J. Chromatogr.*, 1987, **392**, 175.
138. J. M. Levy and W. M. Ritchey, *J. Chromatogr. Sci.*, 1986, **24**, 242.
139. C. R. Yonker and R. D. Smith, *J. Chromatogr.*, 1986, **361**, 25.
140. C. R. Yonker, D. G. McMinn, B. W. Wright, and R. D. Smith, *J. Chromatogr.*, 1987, **396**, 19.
141. A. L. Blilie and T. Greibrokk, *Anal. Chem.*, 1985, **57**, 2239.
142. M. Novotny and P. David, *J. High Resolut. Chromatogr., Chromatogr. Commun.*, 1986, **9**, 647.
143. Y. Hirata and F. Nakata, *J. Chromatogr.*, 1984, **295**, 315.
144. Y. Hirata, F. Nakata, and M. Kawasaki, *J. High Resolut. Chromatogr., Chromatogr. Commun.*, 1986, **9**, 633.
145. E. Lundanes and T. Greibrokk, *J. Chromatogr.*, 1985, **349**, 439.
146. S. B. Hawthorne and D. J. Miller, *J. Chromatogr.*, 1987, **388**, 397.
147. J. A. Graham and L. B. Rogers, *J. Chromatogr. Sci.*, 1980, **18**, 75.
148. S. R. Lipsky and M. L. Duffy, *LC & GC*, 1986, **4**, 898.
149. T. L. Chester, *J. Chromatogr. Sci.*, 1986, **24**, 226.
150. P. J. Schoenmakers, *Lab. News*, 1987, (July 10), 12.
151. D. J. Hawkes, *Int. Labmate Guide*, 1987, p. 15.

CHAPTER 3

Selection of Conditions for an SFC Separation

DIETGER LEYENDECKER

1 Introduction

One of the key features of supercritical fluid chromatography (SFC) is its flexibility. Optimisation of a fluid chromatographic separation can be obtained by changing a variety of parameters. Analyses in gas chromatography (GC) are influenced by changing or programming the analysis temperature and by choosing a suitable stationary phase. In modern GC, capillary columns are preferred. On the other hand, in liquid chromatography (HPLC, GPC, TLC, IC) the variation and programming of the eluent composition are the main tools for optimisation besides the choice of a stationary phase, which is packed into a column as particles. In SFC, however, all the parameters mentioned above bear an influence on separation, *i.e.*, selection of a stationary phase (either in a capillary or in a packed column), selection and/or programming of the mobile phase, and optimisation of the analysis temperature. Additionally, variation of the eluent density and the working pressure are of great importance in optimising SFC separations. Eluent composition, temperature, pressure, and density may be varied singularly or together. This can be achieved by superimposing a temperature programme onto an eluent composition or a density programme. Effective use of all programming techniques requires knowledge of the dependences of chromatographic parameters on physical or technical parameters. In the following sections, influence of all the parameters mentioned above on retention, selectivity, and resolution will be discussed in detail using real chromatograms for illustration.

2 Chromatographic Parameters

To describe the influence of various parameters on a separation, knowledge of some chromatographic parameters may be useful. The capacity ratio of a substrate, k', is calculated from its retention time, t_R, and the dead time, t_0, measured by an inert substrate:

$$k' = (t_R - t_0)/t_0 \tag{1}$$

The ratio of the capacity ratios of two neighbouring peaks, k'_1 and k'_2 ($k'_2 > k'_1$), defines the selectivity, α, between those substrates:

$$\alpha = k'_2/k'_1 \tag{2}$$

The retention time and the selectivity only provide information about the position of the peaks in the chromatogram. They don't say anything about the peak shapes and symmetry. Therefore, a distinct peak pair may show the same selectivity, whether the peaks are well resolved or not. This additional information is included in the plate number, n, and the plate height, h, where:

$$n = 16(t_R/w_b)^2 \tag{3}$$

w_b = distance between the tangents to the peak, measured at the baseline,

$$h = L/n \tag{4}$$

L = column length

Plate number, n, and plate height, h, describe the peak distortion during the time a substrate remains in the separation column. The distortion depends on the linear velocity of the eluent and may be described by the van Deemter equation:

$$h = A + B/u + (C_m + C_s) \times u \tag{5}$$

The A-term is caused by the non-uniformity of flow in a packed column, the B-term by the longitudinal diffusion in the column, and the C_M- and the C_S-terms by the inhibition of diffusion in the mobile and in the stationary phases, respectively. The average linear velocity is given by u. The resolution, R_s, between two neighbouring peaks is usually calculated by:

$$R_s = 2(t_{R2} - t_{R1})/(w_{b1} + w_{b2}) \tag{6}$$

Schmitz[1] derived a modified equation to calculate resolution values, which is also valid for poorly resolved peak pairs, *i.e.*, the baseline peak widths, w_b, are not measurable. Because resolution, R_s, depends on the same variables as the capacity ratio, k', the selectivity, α, and the plate numbers, n, it can be expressed by means of k', α, and n:

$$R_s = \frac{1}{4} \frac{(\alpha - 1)}{\alpha} \frac{k'}{(1 + k')} \sqrt{n} \tag{7}$$

3 Variation of One Parameter

Density

The density of the mobile phase is the most important parameter to influence and optimise separations in supercritical fluid chromatography. Density programming during an analytical run is as common in SFC as temperature programming in GC or programming of eluent composition in HPLC. The influence of density on the solvent properties can be demonstrated using the concept of solubility parameters, first introduced by Hildebrandt and Scott,[2] and extended to supercritical phases by Giddings.[3,4] Following this concept, the solubility parameter, δ, of a solvent depends on its density, ρ:

$$\delta = 1.25 \; P_C^{\frac{1}{2}} \; (\rho/\rho_{liq.}) \qquad (8)$$

P_C : critical pressure

$\rho_{liq.}$: reference density in the liquid state

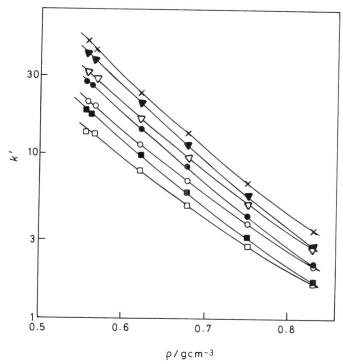

Figure 1 *Capacity ratio* (k')*-density* (ρ) *isotherms for fatty acid phenyl esters and fatty alcohol phenyl ethers. Stationary phase: Spherisorb ODS2; mobile phase:* CO$_2$; T = 39.5 °C; X : n-C$_{18}$H$_{37}$OC$_6$H$_5$; ▼ : n-C$_{17}$H$_{35}$COOC$_6$H$_5$; ▽ : n-C$_{16}$H$_{33}$OC$_6$H$_5$; ● : n-C$_{15}$H$_{31}$COOC$_6$H$_5$; ○ : n-C$_{14}$H$_{29}$OC$_6$H$_5$; ■ : n-C$_{13}$H$_{27}$COOC$_6$H$_5$; □ : n-C$_{12}$H$_{25}$OC$_6$H$_5$
(Reproduced by permission from Ber. Bunsenges. Phys. Chem., 1984, **88**, 841)[5]

The solubility parameter δ varies from 0 up to liquid-like values of 10 at high densities. To solubilise a substrate, the solubility parameters of the substrate and of the solvent should be nearly equal. At low eluent densities, solubility parameters of the mobile phases in SFC are low compared to those of the substrates, but they increase with increasing densities. Therefore, capacity ratios, k', decrease at higher densities (Figure 1)[5] indicating higher solvent strength. The slope of the decrease is not the same for different classes of compounds. The fatty acid phenyl esters and fatty alcohol phenyl ethers used to derive Figure 1, show different k' values at low densities, but some of them fall together at high densities. This means at the same time that selectivity, defined by the ratio of two capacity ratios (Equation 2), decreases at higher densities.

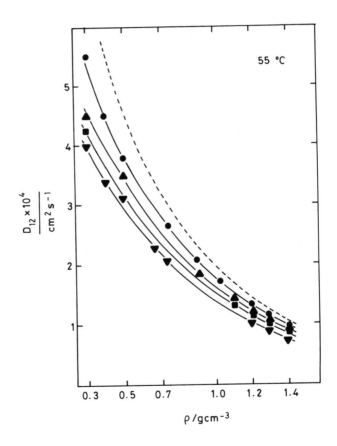

Figure 2 *Binary diffusion coefficient,* D_{12}*, of* ●: *benzene,* ▲: *methyl benzene,* ■: *1,4-dimethylbenzene, and* ▼: *tetrachloromethane, and self diffusion coefficient,* D_{11}*, (----) in supercritical* SF_6 *at 55 °C as a function of density*
(Reproduced by permission from *Ber. Bunsenges. Phys. Chem.*, 1984, **88**, 841)[5]

Looking at separation efficiencies—additionally to selectivities—similar observations were made. Due to lower mass transport rates at higher densities, diffusion coefficients of the substrates and the self diffusion coefficient of the mobile phase also decrease (Figure 2)[5] leading to peak broadening. Assuming constant retention behaviour at the same time, *i.e.*, k' = constant, this results in van Deemter plots of *h versus u*, the increase of which after passing the typical minimum is less steep at lower densities than at high densities (Figure 3). The three different densities represent the elution densities of the aromatic substrates at the same capacity ratio (k' = 2). Although the *h* values at the minimum are the same for all densities, this observation is important for SFC separations. Usually, they are not performed at the optimum linear velocity, *i.e.*, where the minimum in the van Deemter plot occurs, but at 5–10 times faster linear velocities. Here, the loss in efficiency at higher densities is obvious.

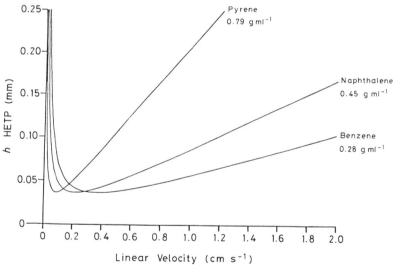

Figure 3 *Effect of carbon dioxide density on the optimum linear velocity, u_{opt}, in a van Deemter plot. The density for each aromatic hydrocarbon was selected to the value given so that each compound had the same capacity ratio (k' = 2)*

In practice, highest resolution is observed in the first part of density-programmed chromatograms. Nevertheless, the possibility of combining high efficiency with high solvating power in one run, makes density programming extremely powerful for resolving complex mixtures or mixtures which contain volatile and hardly volatile components at the same time. For those mixtures, besides linear density programming, asymptotic programs have been used successfully.[6] Additionally, combining density programming with temperature programming to enhance diffusion coefficients during the analysis leads to chromatographic separations, which cannot be performed by any other chromatographic method (see Section 4).

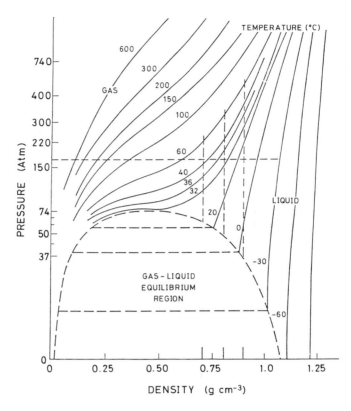

Figure 4 *Dependence of carbon dioxide density on pressure at different temperatures*

Pressure

The influence of column pressure on separations in supercritical fluid chromatography is similar to that of density. Pressure, P, and density, ρ, are related to each other by the temperature, T, as a parameter in a non-linear way by an equation of state. A typical relationship between P, ρ, and T is shown in Figure 4 for carbon dioxide as the mobile phase. Carbon dioxide has a critical temperature of 31.3° C, a critical pressure of 72.9 atm, and a critical density of 0.47 g ml^{-1}. The shape of the P(ρ)-isotherms is more or less sigmoidal. At very high and very low temperatures, the curves become more linear. On the other hand, near the critical point, the slope of the isotherms is very low, *i.e.*, small variations in pressure lead to huge density changes. Exactly at the critical point, the P(ρ)-isotherm has a horizontal tangent. Because of the non-linear relationship between pressure and density, linear density programmes at constant temperature imply non-linear pressure ramps and the other way round. To calculate density programmes from pressure and temperature values, the help of a computer may be useful.

Increasing pressure may show a similar effect on capacity ratios, k', as does increasing density (Figure 5).[7] Looking at pressure regions well above critical, k' values decrease more or less significantly with increasing pressure.

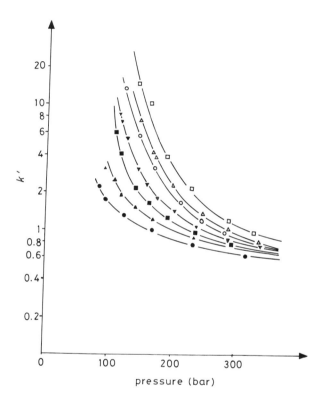

Figure 5 *Dependence of capacity ratios of phenanthrene on pressure at different temperatures:* ●: 300 K, ▲: 310 K, ■: 320 K, ▼: 330 K, ○: 340 K, △: 350 K, □: 360 K. *Eluent:* CO_2; *column: silica*
(Reproduced by permission from *Analusis*, 1985, **13**, 299)[7]

Usually, resolution also decreases with increasing working pressure. This is illustrated in Figure 6 where all fatty acid phenyl ester and fatty alcohol phenyl ether analytes are well separated at low pressure but incompletely resolved at high pressure.[5]

The strong and regular dependence of retention and resolution on pressure at supercritical conditions is utilized by applying pressure programmes to SFC separations to elute volatile components as well as non-volatile ones. At high pressure programming rates, very fast separations can be observed (Figure 7)[10] at the expense of resolution, of course.

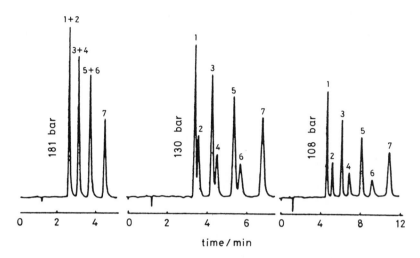

Figure 6 *Effect of pressure on the separation of fatty acid phenyl esters and fatty alcohol phenyl ethers* (1 : *dodecyl phenyl ether;* 2 : *phenyl myristate;* 3 : *tetradecyl phenyl ether;* 4 : *phenyl palmitate;* 5 : *hexadecyl phenyl ether;* 6 : *phenyl stearate;* 7 : *octadecyl phenyl ether). Eluent:* CO_2; *column : ODS* (Reproduced by permission from *Ber. Bunsenges Phys. Chem.*, 1984, **88**, 841)[5]

Some authors reported a maximum in k' [8] and resolution[9] at a particular column pressure using packed columns, when they included experiments at subcritical pressures. This behaviour was related to both pressure-dependent physical and chemical effects. As viscosity increases with higher pressures (Figure 8),[11] the pressure drop along the column is more pronounced at higher pressures. This has to be taken into account, especially when calculating thermodynamic data as a function of column pressure. To describe the flow conditions in a packed column, often the Reynolds number, Re, is used:

$$Re = \rho/\eta \times u \times d_p \qquad (9)$$

ρ = mobile phase density
η = mobile phase viscosity
u = mobile phase linear velocity
d_p = average particle diameter of the packed bed

Following Equation 9, Re depends on the density–viscosity ratio, ρ/η. Plotting ρ/η against column pressure, results in Figure 9.[11] Passing the critical pressure, flow conditions described by the Reynolds number change drastically before reaching a plateau. This may cause irregularities in retention behaviour when separations are carried out in the critical pressure region.

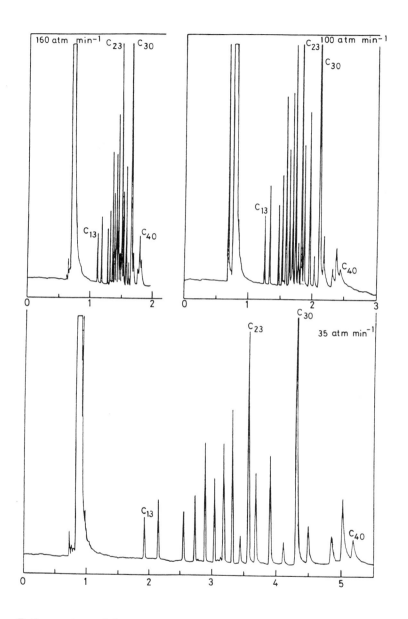

Figure 7 *Comparison of the separations of an n-alkane mixture with pressure ramp rates of 35, 100, and 160* atm min^{-1}. *Temperature:* 100 °C; *column: 5% phenyl polymethyl siloxane*
(Reprinted with permission from *Anal. Chem.*, 1985, **57**, 2629. Copyright 1985, American Chemical Society ©)[10]

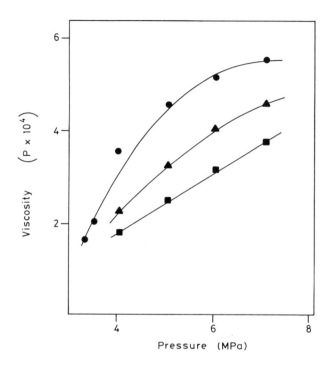

Figure 8 *Viscosity of pentane as a function of pressure at ● : 205 °C, ▲ : 221 °C, and*
■ : 239 °C
(Reproduced from *J. Chromatogr. Sci.*, 1980, **18**, 75, by permission of
Preston Publications, a Division of Preston Industries, Inc.)[11]

Temperature

The discussion about the influence of the temperature, the third physical
parameter besides pressure and density, has to be divided into two parts:
temperature dependences at constant density and at constant pressure.
Changing the temperature at a constant pressure causes variations in the
eluent density.

At constant density, dependence of capacity ratios, k', on temperature can
be calculated by a thermodynamic equation,[12] the Van't Hoff equation.
Assuming that both the phase ratio between mobile phase and stationary
phase and the entropy for the transition of the solutes between the mobile
and the stationary phase are independent of temperature, Equation 10 is
valid:

$$\left(\frac{d \ln k'}{d \, 1/T}\right)_\rho = - \frac{\Delta H_T^0}{R} \qquad (10)$$

ΔH_T^0 = enthalpy for the transition of a solute between the
          mobile and the stationary phases

R   = gas constant

Plotting $\ln k'$ *versus* $1/T$ results in straight lines (Figure 10).[13] The slope of these Van't Hoff plots represents the transition enthalpy, ΔH_T^0.

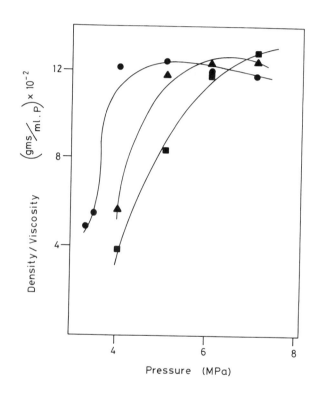

Figure 9 *Ratio of density to viscosity, ρ/η, of pentane as a function of pressure. (For symbols, see Figure 8)*
(Reproduced from *J. Chromatogr. Sci.*, 1980, **18**, 75, by permission of Preston Publications, a Division of Preston Industries, Inc.)[11]

The temperature-dependent behaviour of retention and resolution at constant pressure cannot be described in such a simple manner. In 1967, Sie and Rijnders reported a maximum in capacity ratios, k', at temperatures above critical using n-pentane as the eluent and dialkylphthalates as the substrates (Figure 11).[14] Similar observations were made when analysing

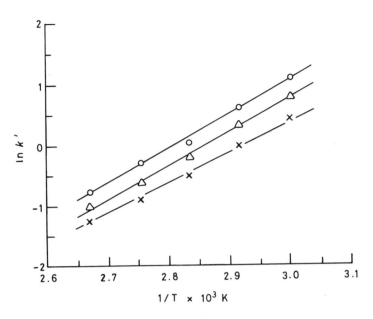

Figure 10 *Van't Hoff plot of* ×: *heptadecane* (C$_{17}$), △: *octadecane* (C$_{18}$), ○:
nonadecane (C$_{19}$). *Eluent:* CO$_2$; *column OV 17*
(Reprinted with permission from *J. Phys. Chem.*, 1985, **89**, 5526. Copyright 1985 American Chemical Society ©)[13]

metal acetylacetonates with carbon dioxide.[15] More detailed experiments covering an extended temperature range from sub- to supercritical conditions resulted in a general shape of the k' *vs.* T-plots[9,16-18] shown in Figure 12 for N$_2$O as the eluent and four aromatic hydrocarbons as substrates. This shape is valid for capillary columns as well as for packed columns, for different substrates and a variety of mobile phases. On increasing the temperature from ambient, *i.e.*, in the liquid state for most eluents, k' values decrease and reach a minimum, thus reflecting increased substrate solubility at higher temperatures. After passing the boiling temperature of the eluent (at pressures below critical) or the critical temperature (at pressures above critical), the k' values increase significantly. This increase is more pronounced the lower the pressure and the higher the molecular weight of the substrate. At moderately high pressures, the curves pass through a distinct maximum and then decrease to values comparable again to those at low temperatures. This behaviour under sub- or supercritical gaseous conditions is caused by two opposing effects: first, by a decrease in the eluent density, and hence in the substrate solubility on increasing the temperature at constant pressure, and second, by the opposing effect of an increase in the vapour pressure of the substrate and an increase in solubility. The second effect overcompensates for the first, when the temperature is increased beyond a certain level. At high pressures, the density decrease is

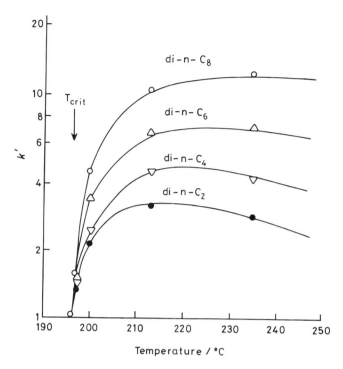

Figure 11 *Effect of temperature on the capacity ratios, k', of di-n-alkylphthalates. Eluent: pentane, pressure 35 kg cm^{-2}, column: PEG 6000 on Sil-O-Cel* (Reproduced by permission from *Sep. Sci.*, 1967, **2**, 729, courtesy of Marcel Dekker, Inc.)[14]

too small for the first effect to become important and, therefore, the k' values do not reach a maximum. Yonker *et al.*[13] evaluated a theoretical model to explain the results mentioned above.

Not only capacity ratios, but also plate numbers, n, increase significantly when the critical temperature of an eluent is exceeded. In Figure 13, three representative chromatograms have been drawn onto a plot of a retention parameter—here elution volume, V_e, of four aromatic hydrocarbons—*versus* temperature at constant pressure in butane.[17] Obviously, peak shapes at comparable elution volumes are better by far for supercritical conditions than for the liquid state. Additionally, selectivity, α, between the homologous substrates used here increases dramatically. According to Equation 7, resolution can be expressed in terms of k', n, and α. Consequently, plots of resolution, R_s, vs. column temperature, T, with CO_2 as the eluent in this case also show maximum values at temperatures somewhat above critical (Figure 14).[17] Figure 15 shows two separations (at 120 °C and 160 °C) of a polyglycol mixture prepared by the reaction of an alcohol with a mixture of ethylene oxide and propylene oxide. Resolution between the different reaction products is much higher at the higher temperature, *i.e.* near the resolution maximum.

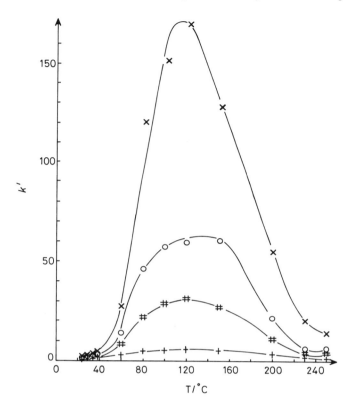

Figure 12 *Dependence of capacity ratios,* k', *vs. temperature,* T. *Eluent:* N₂O;
substrates: +: *naphthalene,* ♯: *anthracene,* ○: *pyrene,* ×: *chrysene; press-*
ure: 120 bar; *column: silica*
(Reproduced by permission from Dietger Leyendecker, Thesis, 1986)[30]

As the mobile phase becomes more gas-like with increasing temperature
at constant pressure, H(u) plots according to the van Deemter equation
(Equation 5) approach those of gases (see Figure 15, Chapter 1). The slope
of the increase in H at higher linear velocities becomes lower at higher
temperatures.

When analysing mixtures of chemically similar components, qualitative
predictions about retention and resolution behaviour may be made with the
knowledge of the relationship mentioned above. On the other hand, on
running samples of different chemical structures, one may get surprising
results. Because the temperature dependence of the retention for each
substrate is different, retention inversions may be observed as is the case in
Figure 16[19] between anthracene and 1,1'-binaphthyl on an alumina column.
Obviously, this enables the analyst to adjust the separation selectivity just by
changing temperature. Without having to lose time by replacing columns or
eluents as in HPLC, optimisation can be achieved very rapidly.

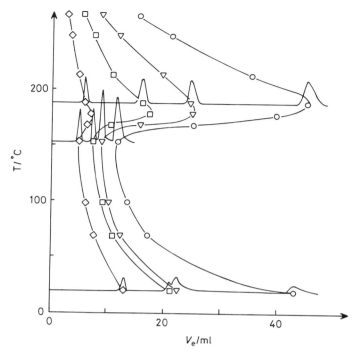

Figure 13 *Dependence of elution volumes,* V_e, *on temperature, T. Eluent: butane; substrates:* $\diamond$: *naphthalene,* $\square$: *anthracene,* $\triangledown$ · *pyrene,* $\bigcirc$: *chrysene; pressure at column exit: 45* bar *column: silica*
(Reproduced by permission from *Ber. Bunsenges. Phys. Chem.*, 1984, **88**, 912.)[17]

Usually, temperature programming in SFC is done by increasing temperature during a pressure, density, or eluent composition programme (see Section 4). In some cases, negative temperature gradients have been applied as a way to increase density (Figure 17).[20] Although density conditions are the same either by decreasing temperature at constant pressure or by increasing pressure at constant temperature, the latter is preferable. The higher diffusion coefficients at the constant higher temperature lead to favourable mass transport properties.

Mobile Phases

In supercritical fluid chromatography, a variety of mobile phases have been used in the past. Physical parameters of selected supercritical fluids are shown in Chapter 1, Table 2.[21-23] From this table it can be seen which temperatures and pressures have to be exceeded to work in the supercritical state. The values for the critical temperatures, T_c, and the critical pressures, P_c, differ significantly for the eluents listed. To analyse thermally labile substrates, one should, of course, choose an eluent with a low critical

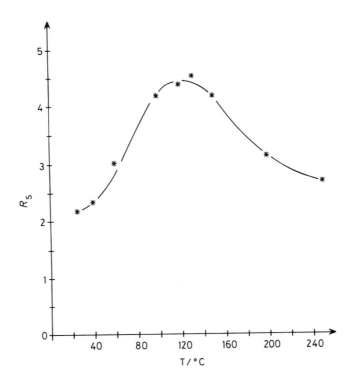

Figure 14 *Dependence of mean resolution, R_s between naphthalene, anthracene, pyrene, and chrysene on temperature, T. Eluent: CO_2; column: silica*
(Reproduced by permission from Dietger Leyendecker, Thesis, 1986)[30]

temperature. With carbon dioxide, nitrous oxide, or trifluoromethane, for example, analyses in SFC can be performed at temperatures not higher than 40 °C which is important for a variety of physiological substances.

To a first approximation, the value of the critical pressure, P_c, is a measure of the solvent strength of a mobile phase. According to the concept of solubility parameters, P_c is related to the solubility parameter by Equation (8). Usually, P_c is higher the more polar the eluent, with some exceptions, *e.g.* xenon.

The critical density, or the density at 400 atm, which is almost the upper pressure limit of most equipment, also gives an idea of the solvent properties. Therefore, supercritical xenon should be an excellent eluent together with the advantages of its transparency in the infrared region of light (Figure 18).[24] Unfortunately, its price is exorbitantly high.

For several reasons, carbon dioxide has become the eluent of choice during the development of SFC. Its critical temperature is low, whereas its critical pressure and critical density are quite high, promising good solvent properties. It is non-toxic, non-flammable and non-explosive. Purification is easy and cheap which is an important point to consider in industry.

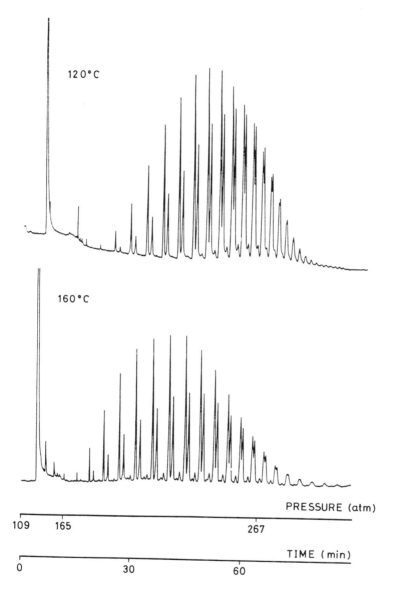

Figure 15 *Pressure-programmed separation of polyglycol oligomers at* 120 °C *and* 160 °C. *Eluent:* CO_2; *column: dimethyl polysiloxane,* 20 m × 50 μm i.d., SE-33, $d_f = 0.50$ μm

However, its greatest advantage is lack of response in the flame ionization detector (FID). Using CO_2 as the mobile phase, universal and sensitive detection becomes possible in SFC. Even in the infrared region, there are some windows in the spectrum of CO_2 where characteristic bands of the analytes remain visible. Therefore, coupling of SFC to a Fourier transform

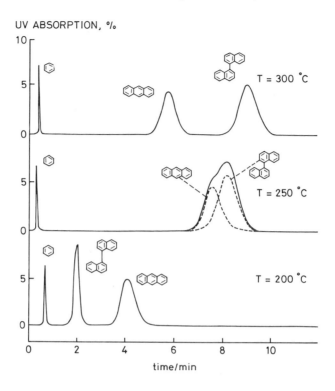

Figure 16 *Effect of temperature on the separation of anthracene from 1,1'-binaphthyl. Eluent: pentane, pressure:* 40 kg cm^{-2}; *column: alumina*
(Reproduced by permission from *Sep. Sci.*, 1967, **2**, 755, courtesy of Marcel Dekker, Inc.)[19]

infrared detector (FTIR), provides the analyst with structural information about the substrates together with their detection. For these reasons, more than 90% of all SFC separations today are performed using carbon dioxide as the eluent.

Within a homologous series of alkane eluents, tuning of the solvent properties and of the physical separation conditions are possible. This can be deduced from Figures 19 and 20[17] where the temperature dependences of capacity ratios (Figure 19) and resolution (Figure 20) have been compared for four different alkanes (propane, n-butane, i-butane, and pentane) using polynuclear aromatic substrates. These plots again show maxima in k' and R_s, when exceeding the respective critical temperatures of each of the eluents. The maxima become more distinct with a lower critical temperature and smaller molecular size of the eluent, indicating a decrease in elution power from pentane to propane. The position of the maxima is shifted to lower temperatures using lower alkanes. Therefore, analysis time or resolution, or both, can be optimised in SFC by choice of the eluent. A low

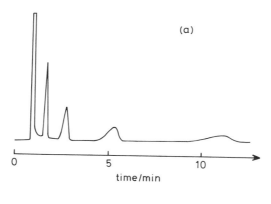

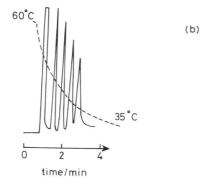

Figure 17 *Separation of* C_{12}-, C_{14}-, C_{16}-, C_{18}-*alkanes.* (a) *Isoconfertic-isothermal 60 °C* (b) *density programmed by decreasing temperature from 60 °C to 30 °C. Eluent: CO_2; column: dimethyl polysiloxane*
(Reproduced by permission from *Fresenius Z. Anal. Chem.*, 1986, **323**, 492)[20]

member of a homologous series, *e.g.*, propane, allows high resolution at a low analysis temperature at the expense of analysis time. On the other hand, simple mixtures of thermally stable components should be analysed using a higher homologue, for example pentane to obtain fast analyses.

The shapes of the k'/T and R_s/T plots seem to be similar for a variety of eluents, *e.g.* Figures 21 and 22[25] for the analysis of polycyclic aromatic hydrocarbons. All of the mobile phases investigated represent eluents with critical temperatures just above ambient. Of course, absolute values of capacity ratios and resolutions depend strongly on the substrates and on the stationary phase. However, trifluoromethane seems to be a much more powerful eluent than ethane, whereas CO_2 and N_2O appear to be quite reasonable eluents.

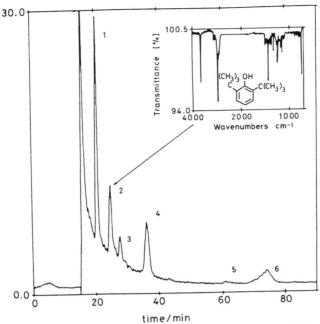

Figure 18 *Gram-Schmidt real-time chromatogram of test mixture* (1: *benzaldehyde*, 2: *2,6-di-t-butylphenol*, 3: *phenol*, 4: *2-naphthaldehyde*, 5: *2-naphthol*, 6: *9-anthraldehyde*). *Eluent: xenon; pressure:* 77 bar; *temperature:* 27 °C (Reprinted with permission from *Anal. Chem.*, 1986, **58**, 164. Copyright 1986 American Chemical Society ©)[24]

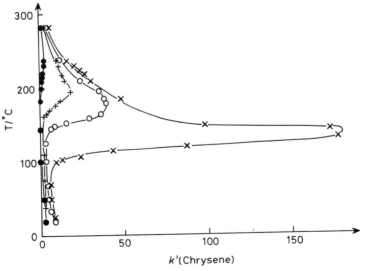

Figure 19 *Comparison of* k' *vs.* T *plots for various alkanes as SFC eluents. Substrate: chrysene. Symbols and pressures:* ●*: pentane,* 36 bar; +: *butane,* 39 bar; ○: *i-butane,* 39 bar; × *propane,* 43 bar. *Column: silica* (Reproduced by permission from *Ber. Bunsenges. Phys. Chem.*, 1984, **88** 912)[17]

The most obvious restriction in SFC today is the lack of a really polar mobile phase. With carbon dioxide, only components with a moderate polarity can be analysed. The use of modifiers, as will be described in the following section, causes some problems with the detectability and the availability of physical data of mixtures.

Supercritical ammonia may solve some of those problems in the future. Although it is quite difficult to handle in terms of stability of apparatus and stationary phases, its solvation power for polar compounds promises a lot. Figure 23 shows the analysis of the hydrochloric salt of a nitrogen base at a very low density.

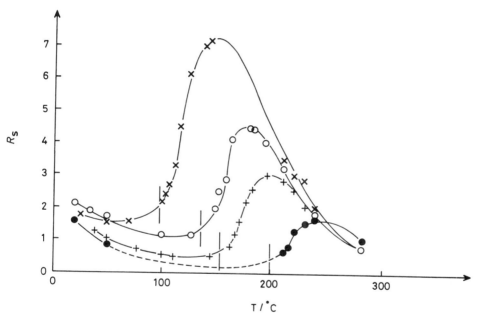

Figure 20 *Temperature dependence of mean chromatographic resolution between naphthalene, anthracene, pyrene, and chrysene for four alkanes as eluents. (For symbols and pressures, see Figure 19). Perpendicular lines indicate the critical temperatures for the alkanes. Column: silica.*
(Reproduced by permission from *Ber. Bunsenges. Phys. Chem.*, 1984, **88**, 912)[17]

Eluent Composition

Continuing the considerations of the previous section, one of the most interesting questions about SFC is how polar can substrates be and still be analysed. This question is somewhat difficult to answer. Using the classification scheme of eluents by Snyder,[26] carbon dioxide shows a polarity

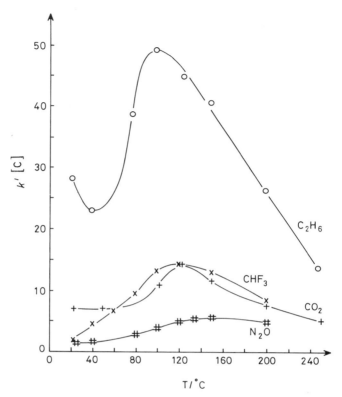

Figure 21 *Dependence of capacity ratios on column temperature at similar reduced pressures, $P_R = P/P_C$. Substrate: chrysene; eluents and pressures: $+$: CO_2, 250 bar, $P_R = 3.39$; $\sharp$: N_2O, 250 bar, $P_R = 3.44$; $\bigcirc$: C_2H_6, 180 bar, $P_R = 3.71$; $\times$: CHF_3, 200 bar, $P_R = 4.13$. Column: silica*
(Reproduced by permission from *J. Liq. Chromatogr.*, 1987, **10**, 1917, courtesy of Marcel Dekker, Inc.)[25]

similar to that of hexane. In practice, CO_2 sometimes behaves like a polar eluent, sometimes as a non-polar eluent. Much more data will have to be collected to obtain more knowledge of what happens in the supercritical state.

The solvent power of the eluents used in SFC may be enhanced by adding a second eluent, the so-called 'modifier' to the basic mobile phase. Separations are often performed by HPLC where the composition of the mobile phase is changed during the run or by adding a modifier before the chromatographic run is started. For programming the eluent composition, a second pump is required. The influence on the retention behaviour of adding a modifier depends on the nature of the substrate, the stationary phase, and on the modifier itself. In SFC using packed columns, strong effects have been observed on addition of small amounts of polar modifiers. This is illustrated by Figure 24 where either a non-polar (hexane) or a polar modifier

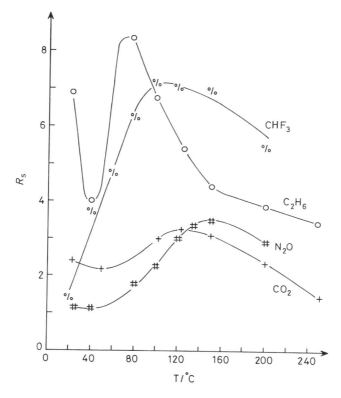

Figure 22 *Dependence of the mean resolution between naphthalene, anthracene, pyrene, and chrysene on column temperature at similar reduced pressures. (For eluents and pressures, see Figure 21). Column: silica*
(Reproduced by permission from *J. Liq. Chromatogr.*, 1987, **10**, 1917, courtesy of Marcel Dekker, Inc.)[25]

(methanol) has been added to CO_2 for the analysis of dimethyl terephthalate on a RP-18 bonded phase column.[27] While no effect was observed on addition of hexane, capacity ratios decreased significantly on adding polar modifiers. This is mainly due to a modification of the stationary phase and not to an enhancement of the solvation power of the mobile phase. On columns packed with silica-based materials, even on bonded stationary phases, there are still some remaining silanol groups. Those groups are mainly responsible for the retention phenomena of polar substrates. The molecules of polar modifiers cover the silanol groups resulting in a more uniform stationary phase in terms of polarity. It was even reported that chemical reactions occurred between silica and 1,4-dioxane used as modifier which modified the stationary phase irreversibly.[28]

If the second eluent component is a worse solvent for the substrates than the basic eluent, capacity ratios increase with higher concentrations of the modifier (Figure 25).[29] While hexane is a good solvent for oligostyrenes, they precipitate on addition of ethanol.

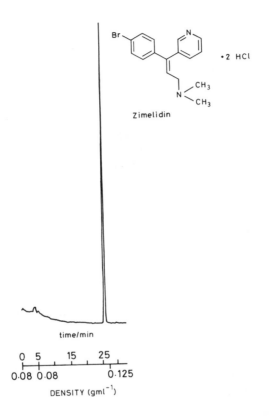

Figure 23 *Chromatogram of Zimelidin with ammonia as the eluent. Temperature: 145 °C, column: 50% nonyl polysiloxane, 10 m × 50 μm i.d., d_f = 0.25 μm, UV detection at 254 nm, 0.2 AU full scale*

Such a basic relationship between retention behaviour and eluent composition is no longer valid when looking at the whole range of concentration. Depending upon column temperature, maxima in capacity ratios of chrysene appeared on addition of 1,4-dioxane to pentane (Figure 26).[36] Although there is no exact explanation for this behaviour, there may be some interaction between the mobile and the stationary phases which causes the substrate to stay longer in the column at certain percentages of modifier.

Due to the higher uniformity of the stationary phase caused by its modification, peaks may be more symmetrical using modified eluents.[28,31] This observation can be expressed in terms of resolution which may reach a composition-dependent maximum (Figure 27).[36] Although capacity ratios decrease on addition of the modifier, this effect is overcompensated for by the decrease in peak widths, and, thereby, an increase in plate numbers. At higher concentrations of modifier, the solvent strength of the mobile phase becomes the determining factor for resolution, and R_s values decrease again. Position and magnitude of the maximum resolution again depend on column temperature.

Modifiers have been applied mainly to elute polar molecules or substrates of high molecular weight. Board *et al.*[27] reported that it was impossible to elute fat-soluble vitamins simply with carbon dioxide, without adding small amounts of methanol. As Figure 28 shows, the higher oligomeric species of a methylphenyl polysiloxane were only eluted with a mixture of 10% ethanol in hexane,[32] even when applying a pressure programme.

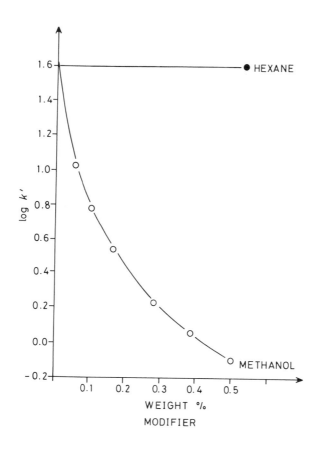

Figure 24 *Effect of modifier concentration on the logarithm of capacity ratios of dimethyl terephthalate. Basic eluent:* CO_2, *column: RP-18*
(Reproduced by permission from Publication No. 43-5953-1647, © 1982, Hewlett-Packard Company)[27]

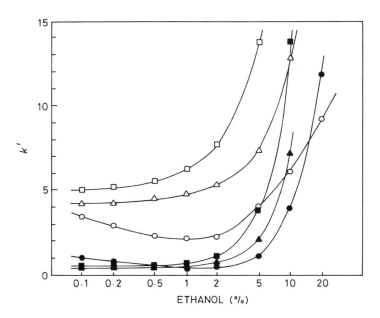

Figure 25 *Dependence of capacity ratios, k', on ethanol concentration for ○: Develosil 100–10, △: Develosil ODS-10, and □: Finesil C_2 columns. Substrate: polystyrene (n = 6), inlet pressure: 50 atm, temperature: 260 °C (closed symbols), 280 °C (open symbols).*
(Reproduced by permission from *J. Chromatogr.* 1984, **315**, 31)[29]

The technique of using eluent composition gradients—such as in HPLC—was introduced to SFC by Klesper, Schmitz, and co-workers. They added a variety of modifiers to alkanes or carbon dioxide and analysed different types of vinyl oligomers. Oligostyrenes, for example, were separated by applying a gradient programme of 1,4-dioxane in propane, butane, pentane, or hexane,[33] *e.g.*, 2-vinyl naphthalene oligomers by 1,4-dioxane/pentane mixtures (Figure 29).[34] Using this method, synthetic mixtures of polystyrene fractions have been the samples of highest molecular weight separated by SFC (Figure 30).[35]

When dealing with the use of modifers, it should be mentioned that some problems arise. By definition, SFC conditions mean pressures and temperatures above the critical values of a certain eluent. Unfortunately, there is no linear connection between the critical data of a binary mixture of eluents and the eluent composition. As Figures 31a and 31b show for the mixtures CO_2/1,4-dioxane, ethane/1,4-dioxane, and pentane/1,4-dioxane,[36] the critical pressures of mixtures, particularly, may display strong non-linearities. Those critical data have been calculated by the method of Chueh and Prausnitz;[37] how far they correspond with experimental data has not yet been measured. Therefore, there is some uncertainty about the physical state of the mobile phase. Additionally, there are very few data about the dependence of density on temperature and pressure which makes density programming using mixed eluents impossible.

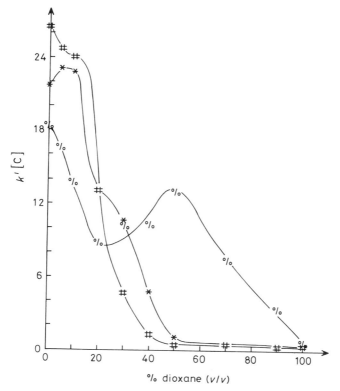

Figure 26 *Dependence of capacity ratios of chrysene,* k' *[C], on eluent composition. Basic eluent: pentane; pressure: 36 bar; temperature: #: 240 °C, * 260 °C, %: 300 °C; column: silica*
(Reproduced by permission from Dagmar Leyendecker, Thesis, 1986).[36]

The second problem with mixed mobile phases is the loss in versatility of detection. Most commercial instruments use the flame ionization detector (FID) as a universal and sensitive detection method. However, as most modifiers are flammable, use of the FID is prohibited. The detector, which has most often been used with mixed eluents, is the UV-detector, requiring substrates containing chromophores. Nevertheless, extending SFC to more polar substrates is one of the main projects of research. New developments in detector technology hold a lot of promise in this field.

Stationary Phases

In SFC, packed columns may be used as well as capillary columns. In the past, there have been some disputes about prefering one type or another. As Schoenmakers explains in the following chapter, both types of columns have their specific benefits and disadvantages. Choosing packed or capillary columns depends only on the analytical problem which has to be solved.

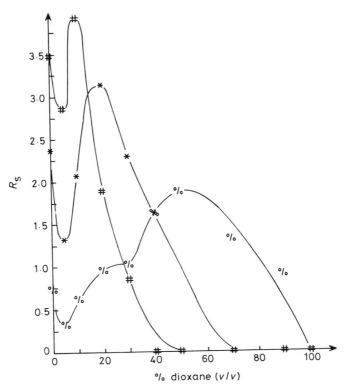

Figure 27 *Dependence of the mean resolution,* R_s, *between naphthalene, anthracene, pyrene, and chrysene, on eluent composition. Basic eluent: pentane; pressure: 36 bar. (Symbols and temperatures as for Figure* 26). *Column: silica* (Reproduced by permission from Dagmar Leyendecker, Thesis, 1986)[36]

Packed columns in SFC are very similar to those used in HPLC. Column diameters vary from 4.6 mm i.d. down to packed capillaries of 320 μm i.d. Packing materials used in the past were bare and modified silica, alumina, and some special materials. Column selectivity follows the same rules as it does in HPLC. Aromatic hydrocarbons, for example, are more retained on an octadecyl silica (ODS) column than on bare silica (Figure 32).[38] Analysis of polar components is difficult on bare silica because of its high retentivity and the limited polarity of pure eluents. Using modifiers may overcome this problem.

A rapid optical resolution of racemic *N*-acetylamino acid t-butyl esters on chiral (*N*-formyl-L-valylamino)propylsilica with methanol-modified carbon dioxide is shown in Figure 33.[39] As this example demonstrates, there is a great variety of different selective stationary phases in packed column SFC.

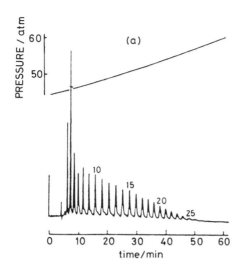

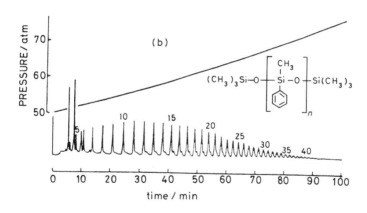

Figure 28 *Separation of OV-17 with pure hexane* (a) *and* 10% *ethanol in hexane* (b). *Column: Develosil 100–10; temperature* 260 °C
(Reproduced by permission from *J. Chromatogr.*, 1984, **315**, 39)[32]

Van Deemter curves in packed column SFC depend on particle diameter in a manner well known from HPLC (Figure 34).[40] Use of smaller particles results in lower values of the flow-dependent minimum in plate heights and a less steep increase in h at higher linear velocities. Therefore, columns packed with small particles show high plate numbers per time unit.

Some features of capillary columns in SFC are quite different from GC columns. Common GC columns cannot be used in SFC, because the stationary phases do not withstand the high solvation power of supercritical eluents. In the literature, separations of GC stationary phases as analytes by several supercritical eluents have been reported.[41,42] To increase the lifetime of the stationary phases in SFC, they have to be cross-linked by radical reactions initiated by peroxides, azo compounds, UV light, or ozone. For GC columns, similar treatments permit higher temperature stability of columns. To avoid uncontrollable retention by remaining silanol groups on the fused silica, SFC columns are very well deactivated before coating with the stationary phase.

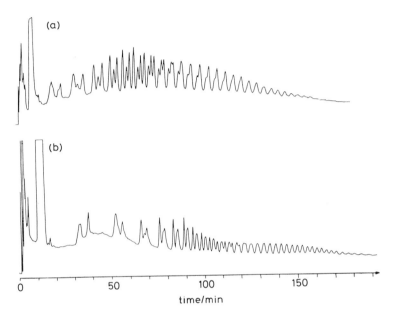

Figure 29 *Separations of 2-vinylnaphthalene oligomers. Column temperatures:* (a) 240°C; (b) 270°C; *gradients:* (a) 5–40% *1,4-dioxane in pentane over* 160 min; (b) 5–50% *dioxane in pentane over* 160 min; *column: silica* (Reproduced by permission from *J. Chromatogr.*, 1983, **346**, 69)[34]

Table 1 *Column efficiencies at* $10\,u_{opt}$ *for different column diameters at* $k' = 3$[a]

d_c (µm)	L (m)	$10\,u_{opt}$ (cm s^{-1})	h (mm)	n	n/m	t_R (min)	n/min
100	24	1.1	0.44	5.4×10^4	2300	145	370
75	24	1.4	0.30	8.0×10^4	3300	114	700
50	23	2.0	0.22	1.0×10^5	4400	77	1300
			(0.19)	(1.2×10^5)	(5200)		(1560)
25	7	4.3	0.18	3.9×10^4	5600	11	3500
			(0.10)	(7.0×10^4)	(10000)		(6400)

CO_2 at 40 °C, 72 atm, 0.22 g ml^{-1}
[a] Values in parentheses calculated from theory

Due to the lower diffusion coefficients in SFC compared to GC, capillary SFC columns should have a smaller internal diameter than GC columns to obtain the minimum in the van Deemter plots at comparable speed of analysis. As a consequence, commercial SFC columns are made with 50 μm and 100 μm i.d. Efficiency in terms of plates per metre becomes higher the

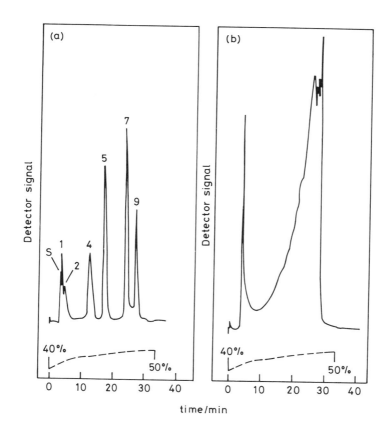

Figure 30 (a) *Chromatogram of a polystyrene mixture. S = solvent; 1: polystyrene sample* (PS) *with number-average molecular weight* $M_n = 36600 \text{ g mol}^{-1}$; *2*: PS *with weight-average molecular weight* $M_w = 10000 \text{ g mol}^{-1}$; *4*: PS *with* $M_w = 93000 \text{ g mol}^{-1}$; *7*: PS *with* $M_w = 254000 \text{ g mol}^{-1}$; *9*: PS *with* $M_w = 600000 \text{ g mol}^{-1}$ *Eluent: 1,4-dioxane in pentane; temperature: 250 °C; pressure at the start of the chromatogram: 185 bar. (b) Chromatogram of an industrial polystyrene sample. Eluent: 1,4-dioxane in pentane; temperature: 250°C; pressure at the start of the chromatogram: 182 bar, columns: silica*
(Reproduced from *Polym. Commun.*, 1983, **24**, 142, by permission of the publishers. Butterworth & Co. (Publishers) Ltd. ©)[35]

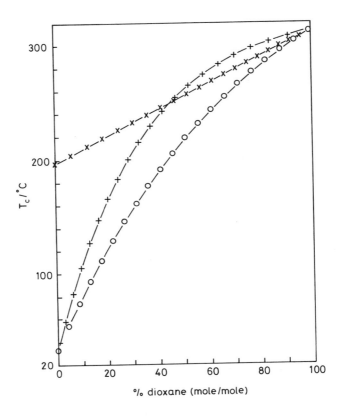

Figure 31a *Dependence of the critical temperature, T_c, on eluent composition. Eluents:*
+: $CO_2/1,4$-dioxane; $\bigcirc$: ethane/1,4-dioxane; $\times$: pentane/1,4-dioxane
(Reproduced by permission from Dagmar Leyendecker, Thesis, 1986)[36]

smaller the internal diameter (Table 1). It is common in SFC to work at 10 times u_{opt} to get faster analyses. The loss in efficiency is not very significant as long as the stationary phase films are thin (Figure 35). If films are no thicker than 1.0 µm, then the loss of efficiency at higher linear velocities is tolerable; common film thicknesses in SFC columns are 0.25 µm and 0.5 µm.

The variation in chemically different stationary phases in capillary SFC is smaller than in packed SFC. Most of them are based on polysiloxanes which are modified by n-octyl-, methyl-, phenyl-, biphenyl-, or cyanopropyl-groups. Following this order, polarity of the stationary phase increases. The selection of a suitable stationary phase follows the same rules as in GC or HPLC, bearing in mind that the frequently-used carbon dioxide is a relatively non-polar eluent. Therefore, non-polar substrates like hydrocarbons are strongly retained by an n-octyl column, whereas free carboxylic acids are retained by a cyanopropyl column. Using a highly polar stationary phase, the range of molecular weights may be extended for the analysis of a

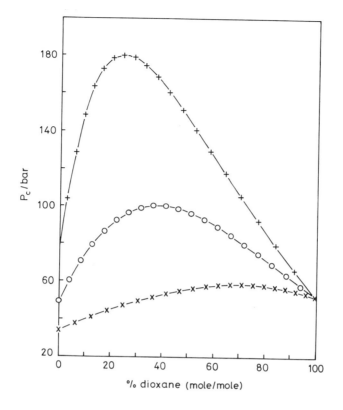

Figure 31b *Dependence of the critical pressure,* P_c, *on eluent composition.* (*Eluents as in Figure* 31a)
(Reproduced by permission from Dagmar Leyendecker, Thesis, 1986)[36]

non-polar mixture. On the other hand, on polar columns, components with only small differences in polarity may be separated. As Figure 36 shows, analysis of a polyglycol mixture on a methyl column only results in separation according to different degrees of oligomerisation, whereas a biphenyl column differentiates distinct species.

4 Variation of Two Parameters

Density/Temperature

As mentioned in Section 3, combined density/temperature programming may be useful to enhance the performance of SFC separations. Increasing temperature during a density programmed analysis, increases volatility of

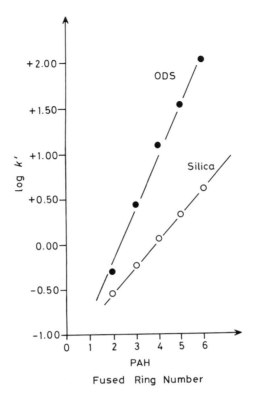

Figure 32 *Logarithm of capacity ratios* vs. *ring number of polycyclic aromatic hydrocarbons for columns packed with silica or ODS-silica material* (Reproduced by permission from Publication Number 43-5953-1690, © 1983, Hewlett-Packard Company)[38]

the substrates and their diffusion coefficients at the same time. Especially for analyses of oligomer mixtures, programming of just one of the parameters results in insufficient separations (Figures 37 and 38). Under those separation conditions, neither using a single temperature programme (Figure 37) nor a density programme (Figure 38) enables the elution of all the components included in a dimethyl polysiloxane sample. Only when both programmes were combined was complete separation of all species up to $n = 70$ observed (Figure 39).

In practice, method development should start using a density programme and subsequently refine this programme. This procedure may result in a separation such as that shown in Figure 40b for a polyglycol sample. Resolution is quite complete in the first part of the chromatogram but seems to be improvable in the high molecular weight region. Superimposing a temperature programme onto the optimised density programme, offers a lot more information about the different oligomeric series (Fig. 40a).

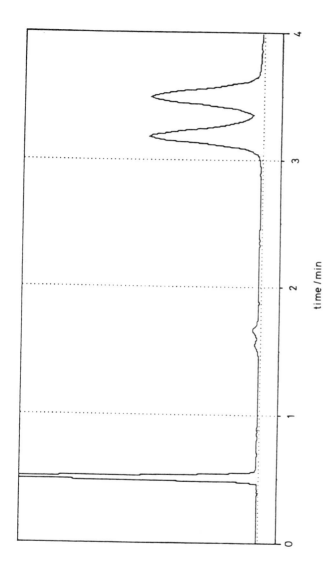

Figure 33 *Optical resolution of the racemic N-acetyl-O-benzyltyrosine t-butyl ester on a chiral stationary phase. Eluent: methanol in* CO_2; *temperature: 60 °C; pressure: 200 bar* (Reproduced by permission from *J. Chromatogr.* 1986, **371**, 153)[39]

time / min

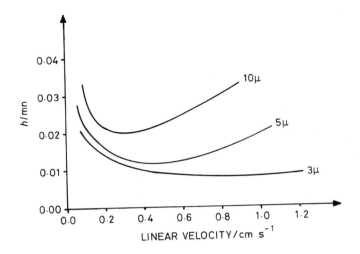

Figure 34 *Influence of particle diameter on column efficiency. Column: ODS Spheri-sorb, eluent:* CO_2
(Reproduced by permission from Publication Number 43-5953-1647, © 1982, Hewlett-Packard Company)[40]

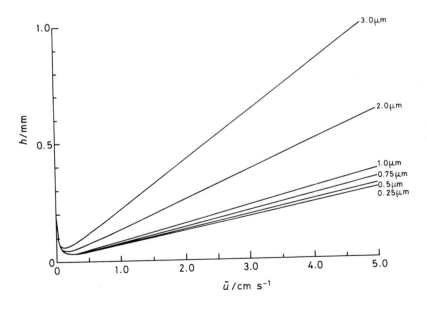

Figure 35 *Effect of film thickness in capillary columns on efficiency. Eluent:* CO_2 *at* 40 °C *and* 72 atm, 50 μm i.d. *columns coated with SE-54* (k' = 1)

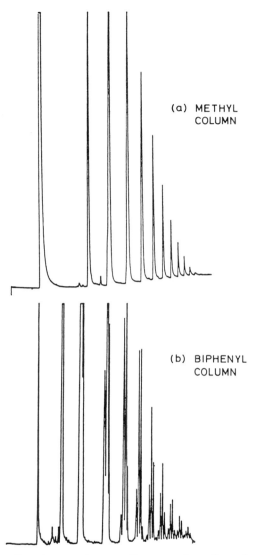

Figure 36 *Effect of stationary phase on the separation of a polyglycol mixture.* (a) *100% methyl polysiloxane;* (b) *50% biphenyl, 50% methyl polysiloxane. Eluent:* CO_2; *temperature:* 100 °C

Pressure/Temperature

The combined influences of column pressure, P, and temperature, T, on retention and resolution have been illustrated using three-dimensional diagrams.[43,44] Either capacity ratios of chrysene (Figure 41) or mean resolution (Figure 42) were plotted *vs.* P and T, using pentane as the eluent and polycyclic hydrocarbons as test substrates.[43]

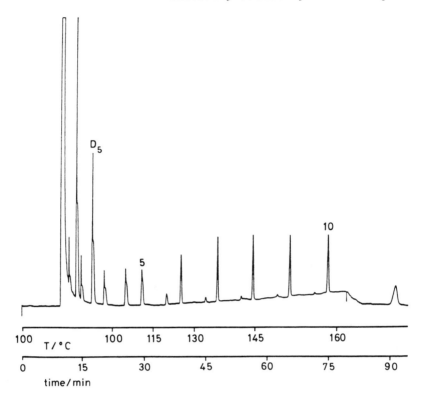

Figure 37 *Isoconfertic separation of dimethyl polysiloxane by temperature program-ming. Eluent:* CO_2 *at* $0.19 \, g \, ml^{-1}$; *column: dimethyl polysiloxane,* $10 \, m \times 50 \, \mu m$; *FID detector*

Comparison of the networks for both k' and R_s shows them to be of similar general shape. The most prominent feature is that of a 'mountain ridge', situated above the critical temperature ($T_C = 196.6°C$) in the super-critical state, where k' or R_s are high. The ridge, *i.e.* k' or R_s, rises with decreasing pressure, and is shifted to lower temperatures. A minimum, or low lying plane is present at lower temperatures in the liquid state. At still lower temperature, *i.e.* near room temperature, k' or R_s rises again.

One may interpolate on the graphs to estimate capacity ratios or resolution for temperature/pressure pairs which lie between the experimental intersec-tions of the isotherms and isobars. Studies on other mobile phases as well as with different modifiers and other non-polar, reasonably volatile substrates, have led to the conclusion that the general shape of the surface of the isocratic networks is at least similar. The absolute values of the k' and R_s, of course, are different, depending on the nature of the mobile and stationary

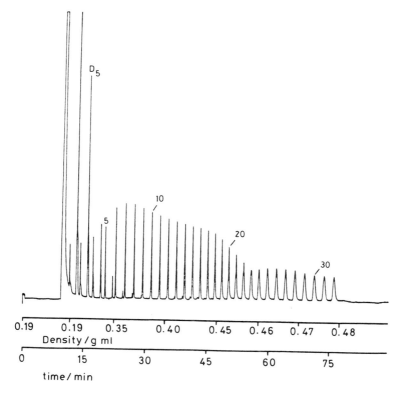

Figure 38 *Isothermal separation of dimethyl polysiloxane by density programming. Eluent:* CO_2 *at 100 °C; column: dimethyl polysiloxane, 10 m × 50 μm: FID detector*

phases, and the molecular weight and vapour pressure of the substrate. Assuming the general shape of the network to be known, it may suffice to run only a few chromatograms to determine the approximate location on the net and to determine in what direction temperature and pressure should be changed to improve resolution. On the other hand, for chemically dissimilar components one may anticipate changes in the general shape of the network.

Composition/Temperature

The contemporaneous dependence of chromatographic parameters on eluent composition and temperature can be demonstrated by the same type of three-dimensional plots as were used in the previous section looking at pressure and temperature.

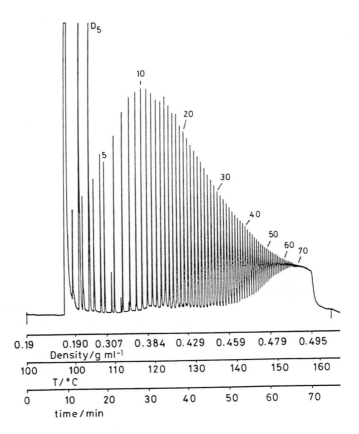

Figure 39 *Separation of dimethyl polysiloxane by combined programming of density and temperature. Eluent:* CO_2*; column: dimethyl polysiloxane,* 10 m × 50 μm; *FID detector*

The dependence on eluent composition and temperature is shown in Figure 43 for the capacity ratios of chrysene, $k'[C]$, and in Figure 44 for the mean resolution, R_s, between four aromatic hydrocarbons using mixtures of pentane and 1,4-dioxane.[45] Usually, k' values decrease when increasing the modifier content in an eluent mixture. There may be ranges in eluent composition, however, where this decrease is reversed to yield a small intermediate rise in k'. In such a range of 1,4-dioxane contents, an increase in 1,4-dioxane leads to an increase in k' instead of the usual decrease. As is described in the literature, not only capacity ratios, but also selectivity and plate numbers show some irregularities when increasing the modifier content.[45] This leads to a complicated shape of the three-dimensional surface in the plots of resolution *vs.* eluent composition and temperature (Figure 44). The general decrease in resolution with increasing modifier concentration is interrupted by several resolution maxima. Therefore, it seems to be very

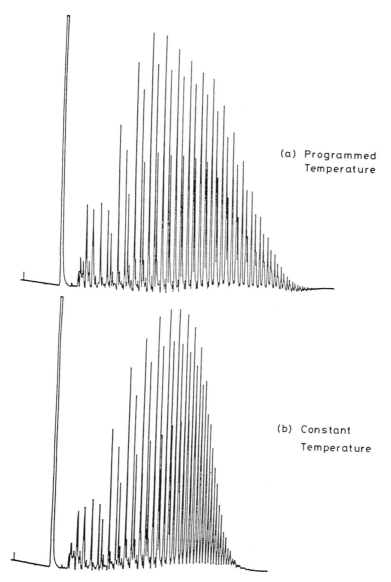

Figure 40 *Density-programmed separation of a polyglycol mixture with programmed (a) and constant (b) temperature. Eluent:* CO_2*; column: dimethyl polysiloxane*

difficult to predict retention and resolution in SFC separation with gradients of eluent composition. As Figure 44 indicates, at constant pressure and composition temperature-dependent resolution maxima have been observed. These maxima shift to higher temperatures with increasing pressure or increasing modifier content. The shift with increasing dioxane concentration

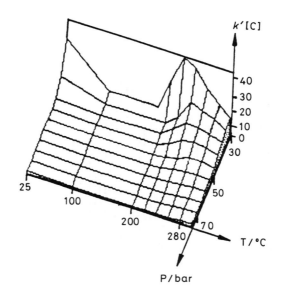

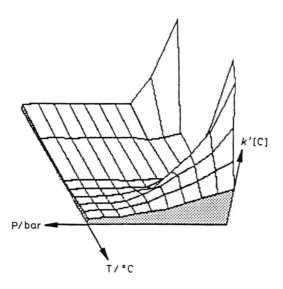

Figure 41 *Capacity ratios of chrysene,* k'[C], *as a function of column pressure, P, and temperature, T. Three-dimensional network shown in two perspectives. Eluent: pentane; column: silica*
(Reproduced by permission from *J. High Resolut. Chromatogr., Chromatogr. Commun.*, 1986, **9**, 566)[43]

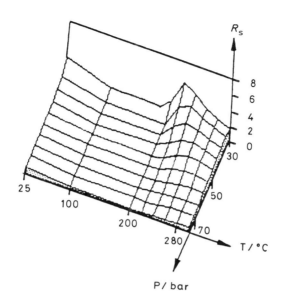

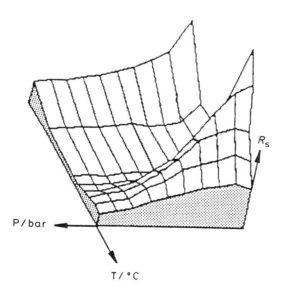

Figure 42 *Mean resolution, R_s, between naphthalene, anthracene, pyrene, and chrysene, as a function of column pressure, P, and temperature, T. Three-dimensional network shown in two perspectives. Eluent: pentane, column: silica*
(Reproduced by permission from *J. High Resolut. Chromatogr., Chromatogr. Commun.*, 1986, **9**, 566)[43]

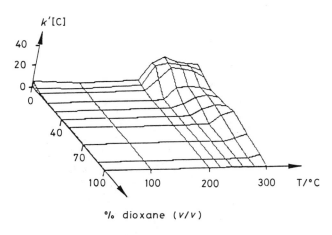

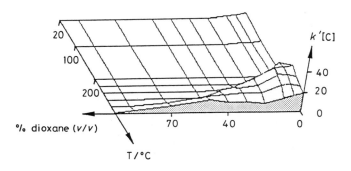

Figure 43 *Dependence of capacity ratios of chrysene, k'[C], on temperature, T, and eluent composition. Three-dimensional network shown in two perspectives. Eluent: 1,4-dioxane in pentane, pressure: 36 bar, column: silica (Reproduced by permission from J. Chromatogr., 1987, **393**, 155)[45]*

is paralleled by the increase in the critical temperature of the binary mobile phase. From chromatographic runs under isobaric, isocratic, and isothermal conditions, triple sets of maximum resolution values, eluent composition and temperature may be evaluated resulting in the plot shown in Figure 45 for the eluent pair carbon dioxide/1,4-dioxane.[46] From this plot, an analysis temperature can be assigned to every eluent composition, resulting in maximum R_s values. In this way, a temperature programme may be evaluated which has to be superimposed onto a gradient of eluent composition. Figures 46a and 46b demonstrate the improvement in the separation of a polystyrene sample using the addition of a temperature programme to an increasing percentage of 1,4-dioxane in carbon dioxide.[46]

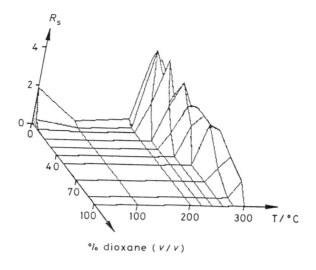

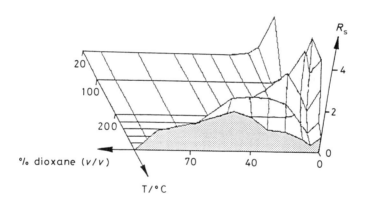

Figure 44 *Dependence of mean resolution,* R_s, *between naphthalene, anthracene, pyrene, and chrysene, on temperature, T, and eluent composition. Three-dimensional network shown in two perspectives. Eluent: 1,4-dioxane in pentane, pressure: 36 bar, column: silica*
(Reproduced by permission from *J. Chromatogr.*, 1987, **393**, 155)[45]

5 Conclusions

In supercritical fluid chromatography (SFC), there are more parameters which influence separations than in any other chromatographic method. Besides variations of stationary and mobile phases, retention, selectivity, and resolution are easily optimised by adjusting temperature, pressure, and density of the mobile phase.

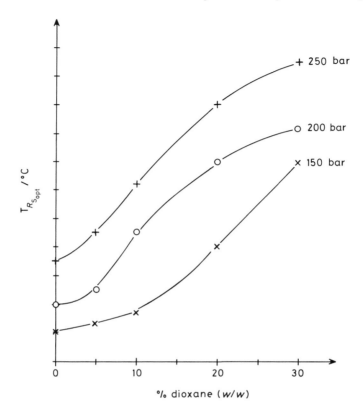

Figure 45 *Temperature of the mean resolution maxima between naphthalene, anthra-*
cene, pyrene, and chrysene, dependent upon eluent composition and column
pressure. Eluent: 1,4-dioxane in CO_2; *column: silica*
(Reproduced by permission from *J. High Resolut. Chromatogr., Chroma-*
togr. Commun., 1986, **9**, 525)[46]

At the present state of development, mobile phase density is the key
parameter in SFC and most separations are performed using variations in
density. Separations are speeded up by applying high densities, while
resolution decreases. Pressure and temperature indirectly imply density
changes. Pressure programming can be used in the same way as density
programming to elute components of high molecular weight or to shorten
analysis times. Temperature is often used as a second parameter to improve
resolution in density- or pressure-programmed SFC analyses. At constant
density, analyses are faster at higher temperatures. At constant pressure,
short analyses are obtained at temperatures slightly below, or well above, the
critical temperature of the eluent. Maxima in retention and resolution are
found at temperatures somewhat above critical. The solvent strength of the
mobile phase can be adjusted either by choosing a suitable eluent or by
adding a second eluent component.

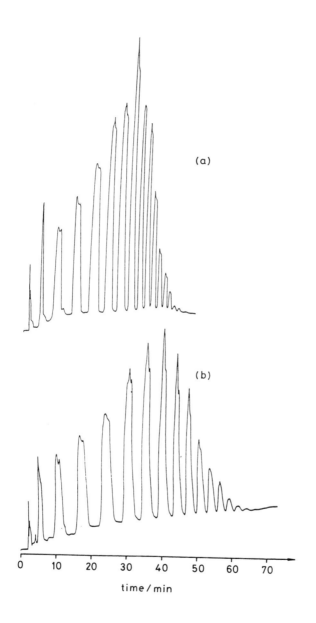

Figure 46 *Eluent composition programmed separations of polystyrene oligomers (PS800) with constant temperature* (a) *and programmed temperature* (b). *Eluent: 1,4-dioxane in* CO_2; *column: silica*
(Reproduced by permission from *J. High Resolut. Chromatogr., Chromatogr. Commun.*, 1986, **9** 525)[46]

Looking to the future, development of techniques using polar eluents seems to be of great importance. The use of other eluents besides carbon dioxide, or addition of polar modifiers will extend the range of applications. The same goal may be reached by developing new stationary phases.

Even with so many parameters to vary, method development in SFC is easy and does not require much time. Once the choice is made of a stationary and a mobile phase which are suitable for the analytical problem, retention as well as selectivity and resolution may be adjusted by varying density, pressure, and/or temperature. Time required for column equilibration is much shorter in SFC using density and temperature programming than in HPLC with eluent gradients. Calculation of density programmes from pressure and temperature values can be achieved comfortably by a personal computer. Therefore, setting up a new method takes only a few minutes, and optimisation of a separation can be done within several hours.

References

1. F. P. Schmitz, *J. Chromatogr.*, 1986, **356**, 261.
2. J. H. Hildebrandt and R. L. Scott, 'The Solubility of Nonelectrolytes', Reinhold, New York, 1950.
3. J. C. Giddings, M. N. Myers, L. McLaren, and R. A. Keller, *Science*, 1968, **162**, 67.
4. J. C. Giddings, M. N. Myers, and J. W. King, *J. Chromatogr. Sci.*, 1969, **7**, 276.
5. G. M. Schneider, *Ber. Bunsenges. Phys. Chem.*, 1984, **88**, 841.
6. J. C. Fjeldsted, W. P. Jackson, P. A. Peaden, and M. L. Lee, *J. Chromatogr. Sci.*, 1983, **21**, 222.
7. P. Mourier, M. Caude, and R. Rosset, *Analusis*, 1985, **13**, 299.
8. B. P. Semonian and L. B. Rogers, *J. Chromatogr. Sci.*, 1978, **16**, 59.
9. D. Leyendecker, F. P. Schmitz, D. Leyendecker, and E. Klesper, *J. Chromatogr.*, 1985, **321**, 273.
10. R. D. Smith, E. G. Chapman, and B. W. Wright, *Anal. Chem.*, 1985, **57**, 2629.
11. J. A. Graham and L. B. Rogers, *J. Chromatogr. Sci.*, 1980, **18**, 75.
12. C. R. Yonker and R. D. Smith, *J. Chromatogr.*, 1986, **351**, 211.
13. C. R. Yonker, B. W. Wright, R. C. Petersen, and R. D. Smith, *J. Phys. Chem.*, 1985, **89**, 5526.
14. S. T. Sie and G. W. A. Rijnders, *Sep. Sci.*, 1967, **2(6)**, 729.
15. F. Bickmann and B. Wenclawiak, *Fresenius Z. Anal. Chem.*, 1985, **320**, 261.
16. M. Novotny, W. Bertsch, and A. Zlatkis, *J. Chromatogr.*, 1971, **61**, 17.
17. F. P. Schmitz, D. Leyendecker, and E. Klesper, *Ber. Bunsenges. Phys. Chem.*, 1984, **88**, 912.
18. D. Leyendecker, F. P. Schmitz, and E. Klesper, *J. Chromatogr.*, 1984, **315**, 19.

19. S. T. Sie and G. W. A. Rijnders, *Sep. Sci.*, 1967, **2**, 755.
20. B. Wenclawiak, *Fresenius Z. Anal. Chem.*, 1986, **323**, 492.
21. 'Matheson Gas Data Book', 1980.
22. 'Handbook of Chemistry and Physics', 60th edn. CRC Press Inc., Boca Raton, 1979.
23. G. N. Lewis and M. Randall, *in* 'Thermodynamics', *eds.* K. S. Pitzer and L. Brewer, McGraw-Hill, New York, 1961, p. 605.
24. S. B. French and M. Novotny, *Anal. Chem.*, 1986, **58**, 164.
25. D. Leyendecker, D. Leyendecker, F. P. Schmitz, and E. Klesper, *J. Liq. Chromatogr.*, 1987, **10**, 1917.
26. L. R. Snyder, *J. Chromatogr.*, 1974, **92**, 223.
27. R. Board, D. McManigill, H. Weaver, and D. Gere, Publication No. 43-5953-1647, Hewlett-Packard Co., Avondale, PA. (1982).
28. F. P. Schmitz, D. Leyendecker, and D. Leyendecker, *J. Chromatogr.*, 1987, **389**, 245.
29. Y. Hirata, *J. Chromatogr.*, 1984, **315**, 31.
30. Dietger Leyendecker, Thesis, 1986.
31. A. L. Blilie and T. Greibrokk, *Anal. Chem.*, 1985, **57**, 2239.
32. Y. Hirata, *J. Chromatogr.*, 1984, **315**, 39.
33. F. P. Schmitz, H. Hilgers, and E. Klesper, *J. Chromatogr.*, 1983, **267**, 267.
34. F. P. Schmitz, H. Hilgers, B. Lorenschat, and E. Klesper, *J. Chromatogr.*, 1985, **346**, 69.
35. F. P. Schmitz and E. Klesper, *Polym. Commun.*, 1983, **24**, 142.
36. Dagmar Leyendecker, Thesis, 1986.
37. P. L. Chueh and J. M. Prausnitz, *AIChE J.* 1967, **13**, 1099, *cited in* R. C. Reid, J. M. Prausnitz, T. K. Sherwook, 'The Properties of Gases and Liquids', 3rd edn., p. 140, McGraw-Hill, New York, 1977.
38. D. R. Gere, Application Note 800-1, Publication No. 43-5953-1690, Hewlett-Packard Co., Avondale, PA. (1983).
39. S. Hara, A. Daobashi, K. Kinoshita, T. Hondo, M. Saito, and M. Senda, *J. Chromatogr.*, 1986, **371**, 153.
40. D. McManigill, R. Board, and D. R. Gere, Publication No. 43-5953-1647, Hewlett-Packard Co., Avondale, PA, p. 21 (1982).
41. Y. Hirata, F. Nakata, and M. Kawasaki, *J. High Resolut. Chromatogr., Chromatogr. Commun.*, 1986, **9**, 633.
42. J. A. Nieman and L. B. Rogers, *Sep. Sci.*, 1975; **10**, 517.
43. D. Leyendecker, D. Leyendecker, F. P. Schmitz, and E. Klesper, *J. High Resolut. Chromatogr., Chromatogr. Commun.*, 1986, **9**, 566.
44. D. Leyendecker, D. Leyendecker, F. P. Schmitz, and E. Klesper, *J. High Resolut. Chromatogr., Chromatogr. Commun.*, 1987, **10**, 141.
45. D. Leyendecker, F. P. Schmitz, D. Leyendecker, and E. Klesper, *J. Chromatogr.*, 1987, **393**, 155.
46. D. Leyendecker, D. Leyendecker, F. P. Schmitz, and E. Klesper, *J. High Resolut. Chromatogr., Chromatogr. Commun.*, 1986, **9**, 525.

CHAPTER 4

Open Columns or Packed Columns for Supercritical Fluid Chromatography— A Comparison

PETER J. SCHOENMAKERS

1 Introduction

Supercritical fluid chromatography (SFC) may either be performed in open (capillary) columns or in packed columns. Over the last few years, this has led to two quite different approaches. Not only are there two different kinds of columns, but the stationary phases, the instrumentation, and the typical applications are also different in the two forms of SFC. Nevertheless, the basic theory of chromatography is equally valid in both situations and this will be the starting point in this chapter for a comparison of the two approaches to SFC.

Some characteristics of the two types of columns, as well as their compatibility with different types of detectors, will be discussed in Section 1.

In Section 2, different characteristics of capillary and packed-column SFC will be discussed from a theoretical point of view, with emphasis on the speed of analysis and on the number of theoretical plates that can be achieved.

Packed columns inherently require a larger pressure drop per unit length, as well as per unit number of theoretical plates. Therefore, the pressure drop over the column is a decisive factor in a comparison between open and packed columns for SFC. Special attention will be paid to column pressure-drop effects in Section 3 of this chapter.

In Section 4 the comparison between capillary and packed-column SFC will be summarised and an attempt is made to identify some of the general application areas for the two different approaches to SFC and to speculate on the future of the technique.

Characteristics of Open Columns

An open column is a narrow, capillary tube, in which a film of a stationary phase is coated on the inside wall. The entire volume of the column, except that occupied by the stationary phase, is available for the flowing mobile phase. For a given stationary phase, only three dimensions are relevant, *i.e.* the inner diameter of the column, the thickness of the stationary phase film and the length of the column.

Table 1 *Effects of the column inner diameter (d_c), the stationary phase film thickness (d_f) and the column length (L) on various parameters in open-tubular chromatography. The power (n) describes the proportionality of the parameters to d_c^n, d_f^n, and L^n. The mobile phase is assumed to be non-compressible and the stationary phase is assumed to behave as a liquid film. In varying the column diameter, the ratio d_c/d_f was assumed constant (i.e. constant reduced film thickness), as was the number of theoretical plates (i.e. L/d_c constant). In varying d_f, or L, the other two parameters were assumed constant. In each case the operating conditions (pressure and temperature) were assumed to be chosen so as to keep the capacity factor constant.*

Parameter	Proportionality power (n)		
	d_c	d_f	L
Column diameter (d_c)	$+1$	0	0
Film thickness (d_f)	$+1$	$+1$	0
Column length (L)	$+1$	0	$+1$
Optimum linear flow-rate	-1	0^a	0
Optimum mass flow-rate	$+1$	0^a	0
Minimum plate height	$+1$	0^a	0
Permissible injection volume (1 plate) V_{inj}	$+3$	0	0
Maximum amount (mass) of sample	$+3$	$+1$	0
Permissible detection volume	$+3$	0	$+\frac{1}{2}$
Concentration in detector	0	$+1$	$-\frac{1}{2}$
Solute mass flow through detector	$+1$	$+1$	$-\frac{1}{2}$
Pressure drop per unit length	-2	0	0
Pressure drop per plate	-1	0	0
Resolution	0	0	$+\frac{1}{2}$
Analysis time	$+2$	0	$+1$

^{a} For thin films, where the plate height is determined by diffusion in the mobile phase. Otherwise, h_{min} will increase with increasing d_f, and u_{opt} will decrease.

The diameter of the column affects the mobile phase flow through the column, the sample loadability, the permissible injection and detection volumes, and the pressure drop over the column. All these effects have consequences for SFC instrumentation. However, changing the column diameter also alters the efficiency (minimum plate height, h_{min}) of an open column, as well as the optimum linear velocity (u_{opt}, the velocity at which $h = h_{min}$). In turn, these parameters affect the required analysis time. Table 1 summarises the effects of changes in the diameter of an open column. The effects in this table are schematic because secondary effects are not taken into account (*e.g.* the mobile phase is assumed to be a non-compressible fluid).

To explain how Table 1 should be read, I will use an example. The table describes the effect of changing the column diameter on the solute mass flow through the detector. If the column inner diameter (d_c) is increased by a factor of 2, and if at the same time we double the film thickness (power $+ 1$ for d_f in the d_c column) and double the column length (power $+ 1$ for L in the d_c column) in order to keep the total number of plates constant, then the injection volume may be increased by a factor 8 (power $+ 3$ for V_{inj} in the d_c column) and the solute mass flow through the detector by a factor 2 (power $+ 1$ in the d_c column).

A simpler effect is that of the column length on the analysis time. Increasing L by a factor of 2, while keeping d_c and d_f constant causes the analysis time to be increased by the same factor 2 (power $+ 1$ for the analysis time in the L column). Because the number of plates has also increased by a factor of 2, the resolution is changed by a factor $\sqrt{2}$ (power $\frac{1}{2}$).

Changing the film thickness of the stationary phase (d_f) also affects a large number of parameters. These effects can be predicted if it is assumed that the stationary phase behaves as liquid, so that molecules are *ab*sorbed in the entire stationary phase film, rather than *ad*sorbed on the interface between the stationary and the mobile phase. Generally, if d_f is increased, we can increase the amount of sample injected in the column, and we can increase the sensitivity. Hence, relatively thick layers of the stationary phase are to be preferred. However, this is only true as long as the stationary phase film thickness does not cause significant losses in the efficiency. Above a certain value, an increase in d_f will cause an increase in h_{min}, and a decrease in u_{opt}, causing an increased analysis time and a reduced sensitivity.

The last column in Table 1 shows the effects of altering the length of the column, with all other factors constant. It is seen that an increase in the length L causes a decrease in sensitivity and an increase in the analysis time. Therefore, the column length should be as short as possible, but sufficient to achieve the required separation. This optimum column length is determined by the number of plates required to achieve separation (n_{ne}) and by the plate height (h). The required number of theoretical plates is determined by the thermodynamics of the system, *i.e.* by the (average) capacity factor of the two solutes to be separated ($\bar{k}'$) and by the ratio between these capacity factors ($\alpha = k'_2/k'_1$). If the required resolution is $R_{s,ne}*$, then n_{ne} can be found from

$$n_{ne} = 4 R_{s,ne}^2 \left[\frac{\alpha + 1}{\alpha - 1} \frac{1 + \bar{k}'}{\bar{k}'} \right]^2 \qquad (1)$$

---

* Typically, for quantitative analysis $R_{s,ne} = 1.5$, whereas for qualitative purposes $R_{s,ne} = 1$ may suffice.

The required column length L_{ne} can be found from:

$$L_{ne} = n_{ne}h \qquad (2)$$

h is determined by the phase system (mobile and stationary phase), the operating conditions (*e.g.* pressure, temperature), and the column diameter. If $L > L_{ne}$, then there is too much separation, less sensitivity, and longer analysis times.

Characteristics of Packed Columns

In packed-column chromatography, the particle size is an additional parameter to affect the observed efficiency and sensitivity. The effects of four relevant parameters in packed-column chromatography are summarised in Table 2.

Table 2 *Effects of the particle size (d_p), the specific surface area (S_a), the column (inner) diameter (d_c), and the column length (L) on various parameters in packed-column chromatography. The mobile phase is assumed to be non-compressible. In varying one parameter the three others were assumed to be constant, except in varying d_p, where the column length was varied in order to keep the number of plates constant. In each case the operating conditions (pressure and temperature) were assumed to be chosen so as to keep the capacity factor constant.*

Parameter	Proportionality power (n)			
	d_p	S_a	d_c	L
Particle size (d_p)	+1	0	0	0
Specific surface area (S_a)	0	+1	0	0
Column diameter (d_c)	0	0	+1	0
Column length (L)	+1	0	0	+1
Optimum linear flow-rate	−1	0	0	0
Optimum mass flow-rate	−1	0	+2	0
Minimum plate height	+1	0	0	0
Permissible injection volume (1 plate)	+1	0	+2	0
Maximum amount (mass) of sample	+1	+1	+2	0
Permissible detection volume	+1	0	+2	$+\frac{1}{2}$
Concentration in detector	0	+1	0	$-\frac{1}{2}$
Solute mass flow through detector	−1	+1	+2	$-\frac{1}{2}$
Pressure drop per unit length	−2	0	0	0
Pressure drop per plate	−1	0	0	0
Resolution	0	0	0	$+\frac{1}{2}$
Analysis time	+2	0	0	+1

From a comparison of Tables 1 and 2 it appears that a number of the effects of the particle size in packed-column chromatography are similar to those of the column diameter in open-tubular chromatography. For example, the optimum linear velocity and the plate height are affected in a similar manner by the two parameters. Other parameters, however, are affected differently, because in packed columns they are determined by both the particle size (d_p) and the column diameter (d_c). In fact, it can be seen from the tables that the effects of changing d_c for an open column are identical to changes in the product $d_c d_p$ in a packed column. Because these latter two parameters can be chosen *independently,* there is an additional degree of freedom. For example, upon reducing the particle size, the permissible detection volume does not decrease if the column diameter is increased at the same time.

It can be seen from Table 2 that the specific surface area affects the maximum permissible amount of sample per unit volume in the column, hence it plays a role similar to that of the film thickness in open columns. If the particles are coated with a film of stationary phase, then an additional parameter d_f (with similar effects as S_a) could be added to the list.

Stationary Phase Effects

Traditionally, the stationary phases used in chromatography are classified according to their physical state. For example, in gas liquid chromatography (GLC) the stationary phase is a liquid, and in gas solid chromatography (GSC) it is a solid adsorbent. In current SFC, however, most of the stationary phases used are neither a solid nor a liquid.

In open-tubular SFC the stationary phase of choice is a polymeric film, which is immobilised by extensive cross-linking, and preferably also by chemical anchoring to the wall. Immobilisation is necessary, because super-critical fluids can be excellent solvents for polymers, especially at higher densities. The main disadvantage of the resulting stationary phases is formed by the relatively low diffusion coefficients. Also, polymers shrink upon cross-linking, so that if the process is taken too far the film may start to crack. Both low diffusion and variations in the film thickness will lead to reduced column efficiencies.

The main advantage of the use of immobilised polymeric films as stationary phases for SFC is their inertness and homogeneity. This allows the elution of fairly polar components as symmetrical peaks with a non-polar mobile phase such as carbon dioxide. Polysiloxane phases show relatively good diffusion characteristics, and, by incorporating different functional groups, different retention and selectivity characteristics may be obtained.[1]

In packed-column SFC, most recent studies have employed packing materials designed for use in high-pressure liquid chromatography. Chemically-bonded stationary phases, which feature a monomolecular layer of organic ligands on a silica surface have been used extensively. Typical reversed-phase materials containing long-chain hydrocarbon groups (*e.g.* octyl, C_8, or octadecyl, C_{18}) have been used, as well as more polar phases containing cyano, amino, or diol groups. With all these phases, however, it appears that the silica surface (containing unreacted or 'residual' silanols) is insufficiently shielded to prevent strong interactions with polar solute molecules. This results in broad, tailing peaks for polar solutes. Generally, chemically-bonded monolayers on silica are not ideally suited for SFC with carbon dioxide as the mobile phase and in subsequent years emphasis is likely to shift towards different classes of phases, for example, polymeric phases or polymers coated on silica.

With respect to surface homogeneity and the elution of polar solutes, polymeric stationary phases (*e.g.* styrene-divinylbenzene copolymers) show better characteristics, but other properties of these phases (efficiency, sample loadability, required mobile-phase density) may be less favourable than those of silica-based phases.

Mobile Phase Modifiers

The poor peak shapes obtained for the elution of polar compounds from chemically modified silicas (HPLC columns) with carbon dioxide as the eluent can be much improved by adding even minor amounts of polar organic solvents to the mobile phase. In analogy with liquid chromatography, such additives are usually referred to as 'modifiers'. Methanol, 2-propanol, and 2-methoxy-ethanol appear to be most popular, but many other solvents (including dioxane, dimethylsulphoxide) have been used.

The addition of organic modifiers has the advantage of allowing the selectivity to be manipulated. However, it complicates the instrumentation, necessitates higher operating temperatures (in order to stay above the critical point of the mixture), and causes increased background and noise with many detector systems. Therefore, improving the characteristics of the stationary phase appears to be a more elegant solution for the elution of polar compounds. Currently, modifiers are more often required in packed-column SFC than in open-tubular SFC.

Detector Considerations

The use of different columns, as well as the choice of the mobile phase and the possible addition of modifiers to the mobile phase will affect the selection of detectors for SFC. A summary of some possible 'universal' detectors is given in Table 3. Packed columns allow a much higher sample loading, and can be adapted to a given detection volume by varying the column diameter without affecting the resolution (see Table 2). Therefore, packed columns are much more readily compatible with ultraviolet (UV) and infrared (IR) detectors than capillary columns. Both flame ionization detectors (FID) and mass spectrometers (MS) are more readily interfaced with open-tubular columns because of the lower flow rates. However, both may also be used in combination with packed columns.

For UV and MS there are few restraints for the selection of the mobile phase. Because of its universality, the FID only allows a very limited number of solvents to be used, carbon dioxide fortunately being one of them. For

SFC-IR, xenon is the ideal solvent (favourable critical parameters, virtually empty spectrum), for those who can afford it. Both CO_2 and SF_6 have large empty 'windows' in their IR spectra. One way to obtain IR spectra over the full range may be to perform subsequent chromatographic separations with two different solvents, *e.g.* CO_2 and SF_6. This is illustrated in Figure 1.

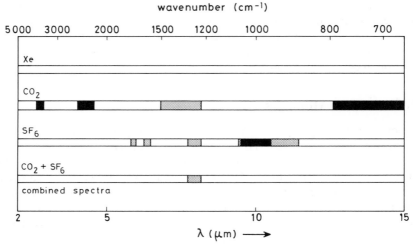

Figure 1 *Schematic illustration of the IR spectra of three solvents for possible use in SFC. Black bars: blocked areas (no IR information). Dashed bars: partially blocked areas (information may be obtained after careful spectral subtraction).[3] The bottom line is a logical or summation of the windows in the spectra of CO_2 and SF_6, showing that a combination of two SFC-IR runs may be used to provide complete IR information*

2 Theory

In this section some of the basic theory of chromatography will be reviewed, from the particular perspective of SFC. The effects of pressure, density, and temperature on the selection and operation of columns are summarised. A simple plate height equation for open-tubular chromatography will be described and used to discuss the required stationary phase film thickness in capillary columns. An equation is derived which describes the required analysis time. This equation is applied to both capillary and packed-column SFC. The maximum number of theoretical plates that can be achieved with a given pressure drop over the column is discussed in the final part of this section.

Effects of Pressure, Density, and Temperature

The density of a supercritical fluid is determined by the pressure and the temperature. In turn, the chromatographically most relevant properties of

Table 3 *Summary of the effects of column and mobile-phase selection on some 'universal' detectors for SFC.*

Detector	Preferred column type	Preferred mobile phases	Modifiers possible?
UV	Packed	UV inactive	yes
FID	Capillary[a]	CO_2, SF_6[b]	no
MS	Capillary[c]	Low-molecular-weight solvents	yes
IR	Packed	Xe, CO_2[d], SF_6[d]	no

[a] Packed columns with $d_c \leqslant 1$ mm possible without splitting.
[b] Gold-plated FID required due to the formation of HF.
[c] Packed columns with $d_c \leqslant 4.6$ mm possible without splitting.
[d] Parts of the spectrum saturated (see Figure 1).

the fluid (viscosity, diffusivity) are largely determined by the density*. The density of supercritical fluids may vary from gas-like to liquid-like values, as can easily be understood from the phase diagram in Figure 2. By passing through the supercritical (SF) region, a gas (G) can be transformed into a liquid (L) by a continuous process, during which the properties of the fluid change regularly. This illustration shows that there is no such thing as a supercritical solvent which combines the viscosity and the diffusivity of a gas with the solvating power (density) of a liquid. Rather, a supercritical fluid

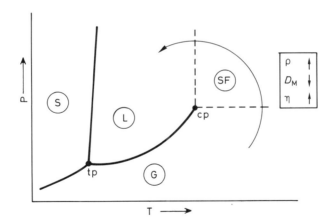

Figure 2 *Phase diagram of a pure substance, illustrating the position of supercritical fluids (SF) between gases (G) and liquids (L). Following the arrow, a gas can be transformed into a liquid by a continuous process, in which the density and the viscosity increase regularly, and the diffusion coefficients decrease correspondingly*

* At constant density ($0.7 \leqslant \rho \leqslant 0.9$ g cm^{-3}) the viscosity of a supercritical fluid has been shown to be independent of the pressure or temperature, whereas at constant densities in the same range the diffusion coefficient for naphthalene in carbon dioxide was shown[2] to vary by no more than 10–20% for temperatures between 0 and 60 °C. At $\rho = 0.6$ g cm^{-3} the variation is 15% between 35 and 60 °C.

can be anywhere in between a gas and a liquid, and in some situations it may be the best compromise between the two. This is schematically illustrated in Figure 3. In this figure the diffusion coefficient is plotted against the density and areas are indicated in which gases, liquids, and supercritical fluids can typically (but not exclusively) be found.

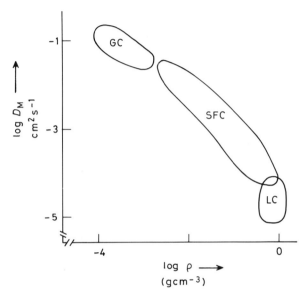

Figure 3 *Schematic diagram illustrating the relationship between the mobile phase density and the diffusion coefficients of solutes. Areas indicate typical gases (G), liquids (L), and supercritical fluids (SF)*

The density (solvating power) required for the elution of samples from a chromatograph is mainly determined by the sample molecules. Volatile solutes can be eluted from a gas chromatograph (GC) because of their volatility. If the solutes are not sufficiently volatile (or not sufficiently stable at the required temperatures), then GC cannot be used. According to Figure 3, we would then either have to refer to liquid chromatography (LC) and sacrifice a typical four orders of magnitude in the diffusion coefficients, or we may use SFC with intermediate mobile phase diffusion coefficients.

A similar schematic diagram may be constructed for the viscosity. However, by applying high pressures (HPLC), the disadvantages of the high viscosities of liquids can be overcome. In the subsequent treatment it will become clear that the magnitude of the diffusion coefficient has a great effect on the efficiency of a given column, on the selection of columns and on the speed of analysis that can be achieved.

The diffusion coefficient is one of the few parameters in chromatography to which the simple rule '*The higher, the better*' can be applied. Therefore separations should ideally be carried out at the lowest possible mobile phase density, which in many cases may imply SFC rather than LC.

Plate Height Equations

The discussion in this section will be based on well known chromatographic theory. The Golay equation is used to describe the relationship between the plate height (h) and the (average) linear velocity of the mobile phase ($\bar{u}$)*, for an open column. The equation reads:

$$h = \frac{2D_M}{\bar{u}} + f(k')\frac{d_c^2\bar{u}}{D_M} + g(k')\frac{d_f^2\bar{u}}{D_S} \tag{3}$$

where (with h in cm and $\bar{u}$ in cm s^{-1}) D_M (cm^2 s^{-1}) is the diffusion coefficient of the solute in the mobile phase and D_S (cm^2 s^{-1}) its diffusivity in the stationary phase, k' the capacity factor (dimensionless retention time), d_c (cm) the column diameter and d_f (cm) the thickness of the stationary phase film on the column wall. $f(k')$ and $g(k')$ are functions of the capacity factor k':

$$f(k') = \frac{1 + 6k' + 11k'^2}{96(1 + k')^2} \tag{4}$$

and

$$g(k') = \frac{2k'}{3(1 + k')^2} \tag{5}$$

In order to compare columns of different diameters, it is very useful to introduce the reduced (dimensionless) plate height (h_r):

$$h_r = h/d_c \tag{6}$$

and the (average) reduced velocity $\bar{v}$:

$$\bar{v} = \frac{\bar{u}d_c}{D_M} \tag{7}$$

into Equation (3). We then find:

$$h_r = \frac{2}{\bar{v}} + f(k')\bar{v} + g(k')\left[\frac{d_f}{d_c}\right]^2\left[\frac{D_M}{D_S}\right]\bar{v} \tag{8}$$

Finally, we may define a (dimensionless) reduced film thickness as:

$$\delta_f = \frac{d_f}{d_c}\sqrt{\frac{D_M}{D_S}} \tag{9}$$

Fields et al.[4] have made a similar substitution by defining a C factor, equal to δ_f^2 in the present notation. The reduced film thickness is a characteristic of

---

* The linear velocity may change along the length of the column if the mobile phase is compressible. The average value is characterized by $\bar{u} = L/t_0$, where t_0 is the residence time of mobile phase molecules in the column.

the column (d_c, d_f, and D_S are parameters of the column and the stationary phase). However, especially in SFC, the mobile phase diffusion coefficient D_M may be strongly affected by the operating conditions. Increasing the density of the mobile phase will tend to decrease the diffusion coefficient. The effect of increasing the temperature (at constant density) is partly cancelled, because D_M and D_S may be affected in a similar way. Most of all, it is important to realise that a column does not show any performance (separation efficiency) of its own. Especially in SFC the observed column efficiency is a strong function of pressure and temperature. Hence, the optimum conditions (optimum flow rate and maximum efficiency) will also be a function of pressure and temperature.

By substituting δ_f into Equation (8) a very simple plate height equation remains:

$$h_r = \frac{2}{\bar{v}} + f(k')\,\bar{v} + g(k')\,\delta_f^2\,\bar{v} \tag{10}$$

Equation (10) expresses the reduced plate height in open-tubular chromatography exclusively in terms of dimensionless parameters.

Packed Columns

For packed columns the situation is somewhat different. There is no analytical *h-u* or *h_r-v* equation. Also, the increase in the plate height with increasing capacity factor is much less pronounced than it is in capillary chromatography. Indeed, this is an advantage of packed columns. To a first approximation, h_r may be assumed to be independent of k', and the Knox equation may be used:*

$$h_r = Av^{\frac{1}{3}} + \frac{B}{v} + Cv \tag{11}$$

In Equation (11) the dimensionless parameters v and h_r for a packed column are defined as:

$$v = \frac{ud_p}{D_M} \tag{7a}$$

where d_p is the particle size and:

$$h_r = h/d_p \tag{6a}$$

Column Pressure Drop

The pressure drop over the column (ΔP) can be found from the Darcy equation, if we neglect the expansion of the mobile phase along the length of

---

* The variation of h_r with k' is small, but not negligible in practice, and functions have been suggested to describe the relationship between the capacity factor and various contributions to the plate height.[5,6]

the column. This approximation is certainly valid for capillary SFC, but also for packed-column SFC not too close to the critical point of the mobile phase.[8] In both cases u is then constant along the column length and:

$$\Delta P = B^0 \eta L u = B^0 \eta N h_r v D_M \qquad (12)$$

If ΔP is in Pa, then the viscosity (η) of the mobile phase may be expressed in Pa.s, the column length (L) in m and the velocity (u) in m s^{-1}. B^0 is the specific permeability coefficient (m^{-2}), which for open columns equals:

$$B^0 = 32/d_c^2 \qquad (13)$$

and for packed columns:*

$$B^0 \simeq 1000 / d_p^2 \qquad (13a)$$

Parameter Estimates

A typical value for the diffusion coefficient in a supercritical mobile phase was taken from reference 4. This 'standard' value is used throughout this chapter, unless the contrary is stated explicitly. The diffusion coefficient of the solute in the mobile phase D_M was taken[4] as 2×10^{-4} cm^2 s^{-1}, and the mobile phase viscosity[2] η as 5×10^{-2} cP (5×10^{-5} Pa.s).

Schwartz et al.[9] have compared the calculated performance of open and packed columns for SFC for three different (hypothetical) solutes, i.e. at three different densities. They ignored the stationary phase contribution to the plate height, which corresponds to assuming $\delta_f = 0$ in the present approach. The conditions of reference 9 are summarised below:

'Small solute' $\rho = 0.28$ g cm^{-3} $D_M = 3.3 \times 10^{-4}$ cm^2 s^{-1}
$\eta = 2 \times 10^{-2}$ cP

'Intermediate size $\rho = 0.54$ g cm^{-3} $D_M = 0.78 \times 10^{-4}$ cm^2 s^{-1}
solute' $\eta = 5 \times 10^{-2}$ cP

'Large solute' $\rho = 0.79$ g cm^{-3} $D_M = 0.30 \times 10^{-4}$ cm^2 s^{-1}
$\eta = 7.2 \times 10^{-2}$ cP

In the terminology of Schwartz, the values assumed by Fields et al.[4] fall somewhere in between those for a small and an intermediate-size solute, i.e. they correspond to a fairly low density, with a rather optimistic estimate for D_M. For most of the present discussion, the estimates of Fields et al.[4] have

---

* For packed-capillary columns, which will be discussed later in this chapter, a more favourable pressure drop is obtained, i.e. approximately: $B^0 \simeq 300/d_p^2$

been used. These will be referred to as 'standard' values. Where appropriate, I will describe different situations based on the estimates of reference 9, and refer to them as small-solute (low density), intermediate-size-solute (medium density), or large-solute (high density) values.

Table 4 *Operating conditions for capillary columns, exclusively in terms of dimensionless (reduced) parameters.* h_r *is the reduced plate height (Equation 6),* $\bar{v}$ *is the (average) reduced velocity (Equation 7), to* k' *is the capacity factor (dimensionless retention time), and* δ_f *is the reduced film thickness (Equation 9).*

$\delta_f = 0.1$ Parameter	$k' = 0$	$k' = 0.5$	$k' = 1$	$k' = 3$	$k' = 10$
$\bar{v}_{opt}$	13.86	7.82	6.42	5.06	4.46
h_{min}	0.29	0.51	0.62	0.79	0.90
h_2	0.36	0.64	0.78	0.99	1.12
h_{10}	1.46	2.58	3.15	3.99	4.53

$\delta_f = 0.3$ Parameter	$k' = 0$	$k' = 0.5$	$k' = 1$	$k' = 3$	$k' = 10$
$\bar{v}_{opt}$	13.86	6.70	5.69	4.77	4.37
h_{min}	0.29	0.60	0.70	0.84	0.92
h_2	0.36	0.75	0.88	1.05	1.15
h_{10}	1.46	3.02	3.55	4.24	4.63

$\delta_f = 1$ Parameter	$k' = 0$	$k' = 0.5$	$k' = 1$	$k' = 3$	$k' = 10$
$\bar{v}_{opt}$	13.86	3.34	3.06	3.15	3.59
h_{min}	0.29	1.20	1.31	1.27	1.11
h_2	0.36	1.50	1.63	1.59	1.39
h_{10}	1.46	6.05	6.60	6.42	5.62

For the coefficients in the Knox equation for packed columns (Equation 11) I used typical values of $A = 1$, $B = 2$, and $C = 0.05$.[7] For packed columns typical operating conditions are assumed to correspond to Equation (11) and to be similar to those encountered in liquid chromatography, *i.e.* either $\bar{v} = 10$, $h_r \simeq 3$, or $\bar{v} = 5$, $h_r \simeq 2$, (Chapter 7 in reference 10). Operating conditions for capillary columns can be calculated from Equation (10). Table 4 lists the calculated values for the optimum reduced velocity ($\bar{v}_{opt}$), the minimum reduced plate height (h_{min} at $\bar{v} = \bar{v}_{opt}$), and the reduced plate heights at $\bar{v} = 2\bar{v}_{opt}$ and $\bar{v} = 10\bar{v}_{opt}$ (h_2 and h_{10}, respectively). These values are listed for three different values of the reduced film thickness (δ_f) and for five different values of the capacity factor k'.

Film Thickness in Open Columns

The introduction of the reduced film thickness δ_f (Equation 9) is very convenient for the comparison of capillary columns for different kinds of chromatography. Such a comparison can be based on equal values of δ_f for the different columns. In Table 5 some possible columns for gas chromatography (GC), SFC and open-tubular liquid chromatography (OTLC) are compared. Three types of GC columns are at the basis of this comparison: a 'standard' column with an i.d. of 250 μm and a film thickness of 0.25 μm, a ('wide bore') thick-film column with an i.d. of 500 μm and a film thickness of 1 μm, and a very thick-film column with an i.d. of 500 μm and a film thickness of 5 μm. All three types of columns are commercially available for GC. Using the estimates for D_M and D_S given in Table 5, the three types of columns represent reduced film thickness of about 0.3, 0.6, and 3. The columns for SFC and OTLC have been calculated for the same values of δ_f.

Table 5 *Column diameters and film thickness for some GC, SFC, and OTLC columns at equal values for the reduced film thickness δ_f.*

GC	$D_M = 5 \times 10^{-2}\,cm^2\,s^{-1}$			$D_S = 5 \times 10^{-7}\,cm^2\,s^{-1}$	
Standard ($\delta_f = 0.32$)		Thick Film ($\delta_f = 0.63$)		Very Thick Film ($\delta_f = 3.2$)	
d_c (μm)	d_f (μm)	d_c (μm)	d_f (μm)	d_c (μm)	d_f (μm)
250	0.25	500	1	500	5
SFC[a]	$D_M = 2 \times 10^{-4}\,cm^2\,s^{-1}$			$D_S = 5 \times 10^{-7}\,cm^2\,s^{-1}$	
Standard ($\delta_f = 0.32$)		Thick Film ($\delta_f = 0.63$)		Very Thick Film ($\delta_f = 3.2$)	
d_c (μm)	d_f (μm)	d_c (μm)	d_f (μm)	d_c (μm)	d_f (μm)
50	0.79	50	1.6	50	7.9
10	0.16	10	0.32	10	1.6
SFC[b]	$D_M = 3 \times 10^{-5}\,cm^2\,s^{-1}$			$D_S = 5 \times 10^{-7}\,cm^2\,s^{-1}$	
Standard ($\delta_f = 0.32$)		Thick Film ($\delta_f = 0.63$)		Very Thick Film ($\delta_f = 3.2$)	
d_c (μm)	d_f (μm)	d_c (μm)	d_f (μm)	d_c (μm)	d_f (μm)
50	2.1	50	4.1	50	20
10	0.41	10	0.83	10	4.1
OTLC	$D_M = 10^{-5}\,cm^2\,s^{-1}$			$D_S = 5 \times 10^{-7}\,cm^2\,s^{-1}$	
Standard ($\delta_f = 0.32$)		Thick Film ($\delta_f = 0.63$)		Very Thick Film ($\delta_f = 3.2$)	
d_c (μm)	d_f (μm)	d_c (μm)	d_f (μm)	d_c (μm)	d_f (μm)
10	0.71	10	1.41	10	—[c]
5	0.35	5	0.71	5	—[c]

[a] 'Standard values'.[4]
[b] High density ('large-molecule values').[9]
[c] Not possible because $d_f \geqslant d_c/2$.

It appears from Table 5 that the film thickness for the conventional 50 μm i.d. capillary SFC columns should be in the order of 1 μm to be comparable to conventional ('standard') GC columns for the application to SFC with low density CO_2 as the mobile phase. This is considerably more than the currently most popular film thickness of 0.25 μm, but it has been shown by

Fields *et al.*[4] that film thicknesses of up to 1 μm are possible with contemporary SFC column technology. However, such a column should not be identified as a thick-film column for SFC. According to Table 5, this term should be saved for columns with film thicknesses of several μm. For applications of capillary columns with high density CO_2 as the mobile phase, it is seen that much thicker films of stationary phase would be desirable, unless the column diameter can be substantially reduced. Not unexpectedly, the high-density mobile phase behaves more similarly to a liquid (*i.e.* OTLC) than to a gas (*i.e.* GC).

The estimated film thickness for OTLC columns is in agreement with current conceptions in this field. Van Vliet[11] has estimated the maximum permissible thickness for OTLC columns based on the condition that the stationary phase contribution to the plate height (third term in Equation 8) should be 20% or less of the mobile phase contribution (second term). For stationary phase diffusion coefficients (D_S) between 5×10^{-8} and 5×10^{-6} cm^2 s^{-1} this resulted in $0.6 < d_f < 2.5$ μm for columns with 10 μm i.d. and $0.2 < d_f < 1.2$ μm for 5 μm i.d. columns. Both the values for the 'standard' and the thick film columns in Table 5 can be found in this range.

The above discussion is based on the assumption that columns with equal reduced film thicknesses should be used. There are several ways to justify this assumption. Basically, the film thickness should be as high as possible, without causing a significantly increased dispersion of the chromatographic peak. When the diffusion coefficient of the solute in the mobile phase (D_M) decreases when going from GC to SFC to LC, the film thickness of a given stationary phase may increase (Equation 9). This is not an uncommon situation, since cross-linked (silicone) polymers may be used in all three situations. Also, it is essential that d_f is increased, as can be demonstrated by expressing the capacity factor in mole fractions as follows:

$$k' = \frac{c_{i,S}}{c_{i,M}} \frac{V_S}{V_M} \simeq \frac{x_{i,S}}{x_{i,M}} \frac{\rho_S}{\rho_M} \frac{M_M}{M_S} \frac{V_S}{V_M} \tag{14}$$

In Equation (14) c is the concentration (mol l^{-1}), V the volume of one of the two phases in the column, x the mole fraction, ρ the density, and M the molecular weight. Subscript S indicates the stationary phase, M the mobile phase, and i the solute.

Equation (14) shows that the distribution coefficient in terms of concentrations ($K_c = c_S/c_M$) can be maintained constant if we keep the phase ratio (V_S/V_M) constant (or approximately if we keep the ratio d_S/d_M constant). However, when going from GC to SFC to OTLC, we greatly increase the density of the mobile phase (ρ_M), and the second part of Equation (14) shows that the distribution coefficient in terms of mole fractions ($K_x = x_S/x_M$) may only be kept constant if the increase in ρ_M is compensated by an increase in V_S. Assuming x_S to be constant (*i.e.* the same stationary phase at the same temperature), x_M will be much lower in SFC and OTLC than it is in GC, unless V_S is increased.

A low value of x_M (a high value of K_x) is highly undesirable in chromatography. In the first place, the column (stationary phase) will rapidly be overloaded. In the second place, the sensitivity of the detection will be very low and the detection limits will be high. These two factors together mean that the dynamic working range (the ratio of the maximum permissible amount of sample and the lowest amount that can be detected) will be very small.

Clearly, the film thickness (d_f) must be chosen as high as possible, and the reduced film thickness δ_f is an excellent indicator for the effect that the stationary phase will have on the peak broadening. It may be concluded from Table 4 that a reduced film thickness of 0.3 causes little additional band broadening (compare h_r values for $\delta_f = 0.1$ and $\delta_f = 0.3$). In other words, the summary of the (beneficial) effects of increasing the film thickness in open-tubular chromatography given in Table 1 is approximately valid for $\delta_f \leqslant 0.3$. When $\delta_f = 1$, however, there is seen to be a considerable effect. Therefore, $\delta_f = 0.3$ seems to be a reasonable value for a 'standard' column.

The data in Table 5 can be used to estimate stationary phase film thicknesses for capillary columns in different kinds of chromatography, provided that stationary phases with similar diffusion coefficients are used. Equation (9) indicates the significance of the stationary phase diffusion coefficient D_S. It is shown that if a stationary phase with improved diffusion characteristics can be used, the film thickness may be increased. For example, a ten-fold increase in D_S allows a three-fold increase in d_f. This calls for stationary phases with a low viscosity (*e.g.* a low degree of cross-linking), yet with a high stability.

From Table 5 it may be concluded that contemporary capillary SFC columns are most useful for separations requiring relatively low mobile-phase densities. For higher densities, either smaller diameters or thicker stationary phase films are required. It will next be shown that the former approach is much to be preferred.

Speed of Analysis

A very simple expression for the analysis time in chromatography can be found from the fundamental retention equation:

$$t_R = t_0(1 + k') = (L/u)(1 + k') \tag{15}$$

where t_R is the retention time of the solute with capacity factor k', and t_0 is the hold-up time of the column. By substituting $L = nh$ and the reduced parameters of Equations (6) and (7), we find:

$$t_R = \frac{nh_r d^2}{\bar{v}D_M}(1 + k') \tag{16}$$

where d is the characteristic dimension ($d = d_c$ for capillary columns, and $d = d_p$ for packed columns).

I will compare capillary and packed volumes for a given separation (given value of n and given value of D_M). Typical values for h_r and $\bar{v}$ for capillary columns may be obtained from Table 4. Based on Table 5 and the discussion above, $\delta_f = 0.3$ will be assumed. For fairly high k' values it appears from Table 4 that $\bar{v} = 45$ and $h_r = 4.5$ are reasonable values (see also reference 4). Using $\bar{v} = 10$ and $h_r = 3$ for packed columns and using Equation (16) twice, we find:

$$\frac{t_{R,c}}{t_{R,p}} = \frac{d_c^2}{3d_p^2} \qquad (17)$$

where $t_{R,c}$ denotes a capillary column and $t_{R,p}$ a packed column. Under these conditions it is seen that, for a given separation, the analysis time for a capillary and a packed column are equal if $d_c = \sqrt{3}d_p$. Therefore, in terms of the required time of analysis for a given separation, a contemporary 50 μm i.d. capillary column corresponds to a packed column with 30 μm particles. Hence, for fast analysis, capillary SFC is not better today than packed-column SFC was 15 years ago. Contemporary packed columns with particle sizes of 10, 5, and even 3 μm correspond to capillary columns of about 17, 9, and 5 μm i.d. If $\bar{v} = 5, h_r = 2$ had been chosen for the conditions of the packed column, the factor would have been 2 instead of $\sqrt{3}$ and the i.d. of capillary columns corresponding to 3 to 10 μm particles would vary between 6 and 20 μm.

The speed of analysis in capillary SFC cannot be increased by increasing the linear velocity ($\bar{v}$), since at high velocities h_r is proportional to $\bar{v}$, and the ratio $h/\bar{v}$ appears in Equation (16). Therefore, in terms of speed analysis, capillary SFC cannot compete with packed-column SFC, unless the column diameters can be greatly reduced.

Maximum Number of Plates

We have seen above that packed columns can provide a given number of plates more rapidly than can capillary columns in SFC. However, this is only true if the required number of plates can indeed be provided. The maximum number of plates that can be achieved is a function of two factors: the pressure drop per theoretical plate and the maximum pressure drop that can be allowed over the column. It is reasonable to assume that the same pressure drop may be allowed over a capillary (c) and a packed column (p), in which case we obtain from Equation (12):

$$[B^0 \eta n h_r \bar{v} D_M]_c = [B^0 \eta n h_r \bar{v} D_M]_p \qquad (18)$$

For capillary columns $B^0 = 32/d_c^2$, $h_r = 4.5$ and $\bar{v} = 45$. For packed columns $B^0 \simeq 1000/d_p^2$, $h_r = 3$ and $\bar{v} = 10$. Since the mobile phase properties are the same on either side of Equation (18), we obtain:

$$n_{c} = 4.63\left[\frac{d_{c}}{d_{p}}\right]^{2}n_{p}. \tag{19}$$

Equation (19) allows a rapid comparison between the maximum numbers of theoretical plates that can be obtained on two different kinds of columns at a given pressure drop. For example, a 50 μm capillary SFC column will allow 116 times as many plates as a column packed with 10 μm particles, and 463 times as many as one with 5 μm particles. Hence, capillary SFC will allow much higher plate numbers to be achieved.

Table 6 *Summary of the (calculated) maximum numbers of theoretical plates and the required analysis times for different kinds of SFC columns. Figures are relative to a 5μm packed column, for which $n_{max} = 20,000$, $t_{0} = 1$ min and $t_{ne} = 5$ min have been assumed[12]. $\bar{v} = 10$, $h_{r} = 3$ for packed columns, $\bar{v} = 45$, $h_{r} = 4.5$ for open columns.*

Column type	Diameter[a] μm	n_{max} [b]	$(t_{ne})_{n=n_{max}}$ [c] min	$(t_{ne})_{n=20,000}$ [d] min
Open	8.5	268,000	64.5	4.82
	17	1,070,000	1030	19.3
	50	9,260,000	77,000	167
	100	37,000,000	1,235,000	667
Packed	3	7,200	0.65	1.80
	5	20,000	5	5
	10	80,000	80	20
Packed-capillary	3	24,000	2.15	1.80
	5	66,700	16.7	5
	10	266,000	266.7	20

[a] d_{c} for open columns, d_{p} for packed columns.
[b] Based on $n_{max} = 20,000$ for a 5 μm packed column.
[c] For $n = n_{max}$ and $k'_{\omega} = 4$, based on $t_{ne} = 5$ min for a 5 μm packed column.
[d] For $n = 20,000$ and $k'_{\omega} = 4$, based on $t_{ne} = 5$ min for a 5 μm packed column.

Table 6 shows a comparison of the maximum numbers of theoretical plates and the required analysis times for open and packed columns of various dimensions. Two different sets of values for required analysis times are shown in the Table. The first one, $(t_{ne})_{n=n_{max}}$, is the time it would take to achieve the maximum number of theoretical plates on each column. Hence, it will take more than a million minutes (and a 10 mile column) to realise the full potential ($n_{max} = 37$ million plates) of a 100 μm i.d. capillary column. The second analysis time $(t_{ne})_{n=20,000}$ corresponds to the realisation of 20,000 theoretical plates. The numbers in the table are based on the assumption that 20,000 theoretical plates can be achieved on a 5 μm packed column within a few minutes, which is a good reflection of the current state of the art in packed-column SFC.[12]

Two effects are strikingly obvious from Table 6: (*a*) the very large numbers of theoretical plates that can be obtained with a capillary column at the same pressure drop required for a packed column, and (*b*) the dramatically long analysis times required. Very high numbers of theoretical plates may be reached in SFC (as indeed in GC and LC) by very patient people. Of more practical significance is the time required for realising a separation requiring 20,000 plates with $k'_\omega = 4$ for the capacity factor of the last peak. Contemporary open columns with diameters of 50 or 100 μm require much longer analysis times than do conventional (HPLC) packed columns. The situation can only be changed by a drastic reduction in the diameter of capillary columns.*

Packed-capillary columns may be of great interest for SFC *if* it turns out that they indeed provide a larger number of theoretical plates (at equal pressure drop and flow rate) in comparison with conventional packed columns. Even if the speed of analysis were to be slightly reduced, as is suggested in reference 13, then they would still compare favourably with contemporary open columns.

3 Packed-column SFC

We have seen in the previous section that, in terms of analysis times, packed-column SFC on HPLC columns (particles of 3 to 10 μm) is far superior to open-tubular SFC on columns of 50 or 100 μm i.d. However, open columns may provide much higher numbers of theoretical plates at the same pressure drop. Therefore, the effects of the pressure drop over the column are a key issue in SFC. The maximum permissible pressure drop will determine when packed columns may be used for rapid analysis and when capillary columns are needed for high efficiency separations. This section begins with a description of the effects of the column pressure drop in packed-column SFC.

Subsequently, a number of instrumental considerations of packed-column SFC and the essential features of programming techniques are briefly reviewed. The potential of packed-capillary columns, which has been touched upon in previous sections, will be summarised.

In the last part of this section an attempt will be made to establish some rules for the selection of packed columns for SFC.

Effects of Column Pressure Drop

In the previous section I have discussed the *relative* numbers of theoretical plates that may be achieved on different columns. Absolute numbers (Table 6) have been related to an estimated number of plates in a particular

---

* The required analysis time for complex samples may be reduced by programmed elution (*e.g.* pressure or density programming), at the expense of reduced resolution. However, as this does not change the hold-up time of the column, only about a factor of 2 can be gained on the numbers in the last column of Table 6.

situation (20,000 plates for a 5 μm packed column with $t_0 = 1$ min). The number of plates that can be achieved will depend on the nature and density of the mobile phase (determining η and D_M in Equation 12) and on the absolute pressure drop that may be allowed over the column.

We have previously studied the effects of the column pressure drop on the observed capacity factor[14] and on the efficiency[12] in packed-column SFC. The flow-rate (linear velocity) will increase with increasing pressure drop. If the density does not vary along the length of the column (*i.e.* if the density is high or the pressure drop low[8]), then $\bar{v}$ is proportional to ΔP. Because Equation (11) shows a minimum value for h_r at an optimum value for $\bar{v}$, there is an optimum value for the pressure drop over the column at which the highest possible number of theoretical plates can be obtained from the column. Working below the minimum is extremely unfavourable, because it implies reduced efficiencies at lower flow-rates, *i.e.* it takes longer to achieve less. Therefore, for a given column, with a given mobile phase at a given pressure and temperature, there is a minimum pressure drop (ΔP_{opt}), for operating the SFC system.

When the pressure drop becomes too high, however, the density at the column outlet (ρ_{out}) may start to decrease. This may lead to increased capacity factors at the column exit and therefore to broader peaks.* When ΔP is increased further, ρ_{out} decreases further, and the solubility of the solute may become a limiting factor. In that case very broad, poorly shaped peaks will be observed in the chromatogram. Clearly, there is a minimum value for ρ_{out}, which means that at a given temperature there is a minimum column outlet pressure. Thus, at a given temperature and a given inlet pressure, there is a maximum permissible pressure drop (ΔP_{max}) over the column.†

Packed columns in SFC should be operated between ΔP_{opt} and ΔP_{max}. Figure 4a shows an example of a plot of the observed number of theoretical plates as a function of the column pressure drop for a 25 cm long column packed with 8 μm particles. Under the conditions of Figure 4 (P = 122 bar, T = 40 °C) and for the three test solutes employed, the optimum pressure drop is seen to be about 20 bar and the corresponding number of plates is in excess of 20,000. At ΔP = 20 bar, $t_0 = 67$ s. Hence, separations with more than 20,000 plates can be achieved in a few minutes.

Figure 4b shows the variation of the observed capacity factor [*i.e.* $k'_{obs} = (t_R - t_0)/t_0$] with the pressure drop over the column. It is seen that the observed capacity factors at the optimum column pressure drop differ significantly (about 20%) from the extrapolated values at ΔP = 0.

---

* The width of a chromatographic peak as it appears on a recorder trace is determined by the volume occupied by the peak just before leaving the column (characterised by the standard deviation in length units σ_L), the capacity factor and the linear velocity at the column outlet (k'_{out} and u_{out}, respectively). In time units we find $\sigma_t = \dfrac{\sigma_L(1 + k'_{out})}{u_{out}}$.

† One way to stay above the minimum value for ρ_{out} is to keep P_{out} constant and to manipulate the flow rate by varying the inlet pressure (P_{in}). However, increasing P_{in} will lead to lower capacity factors and thereby to a loss in resolution.

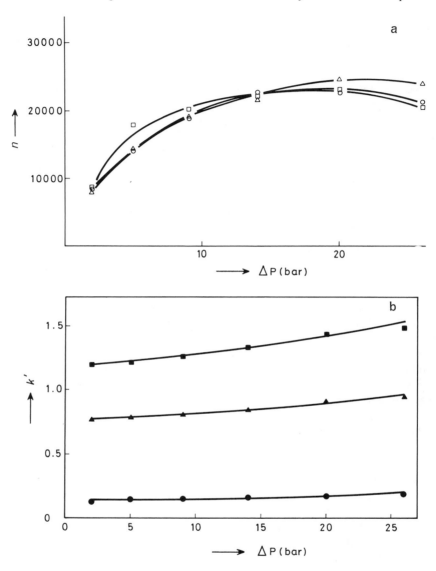

Figure 4 *Variation of* (a) *the number of theoretical plates* (n) *and* (b) *the observed capacity factor* (k') *with the pressure drop over the column* (ΔP). *Column: 25 cm × 4.6 mm i.d., packed with* 8 μm *CP-Spher ODS silica* P_{in} = 122 *bar,* T = 40 °C. *Solutes:* ○, ●: *ethylbenzene,* △, ▲: *naphthalene, and* □, ■: *biphenyl.*
(Reproduced by permission from *Chromatographia,* in print.)[12]

Figure 5 shows quite different results, obtained on another column (25 cm long, packed with 5 μm particles) under the same conditions as Figure 4 (P = 122 bar, T = 40 °C). Thus, the only parameter that has been changed is

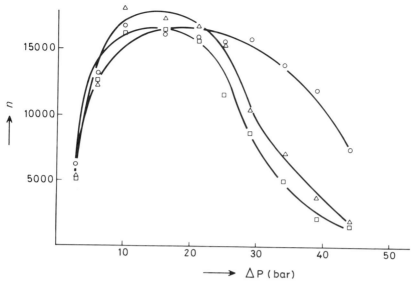

Figure 5 *Variation of the number of theoretical plates (n) with the pressure drop over the column (ΔP). Column: 25 cm × 4.6 mm i.d., packed with 5 μm Rosil ODS silica. Conditions and solutes as in Figure 4.*
(Reproduced by permission from *Chromatographia*, in print.)[12]

the particle diameter. It is shown that the maximum efficiency is lower than on the column packed with 8 μm particles, that the optimum pressure drop is lower, and that the number of plates decreases very rapidly at higher values for ΔP. When extrapolating the results of Figure 4, a much higher optimum pressure drop would have been expected. According to Equation (12), by substituting $n = L/h = L/h_r d_p$ and Equation (13a), we find

$$\Delta P = \frac{1000 \eta L \bar{v}_{opt} D_M}{d_p^3}. \tag{20}$$

With all parameters in the numerator equal for the two columns, we obtain $\Delta P \propto d_p^{-3}$, and thus the optimum pressure drop for the 5 μm column is expected to be $(8/5)^3 \simeq 4$ times higher than for the 8 μm column. This results in an estimate of $\Delta P_{opt} \simeq 4 \times 20 = 80$ bar. Clearly, however, this pressure drop is too high, because it would correspond to a column outlet pressure of about 40 bar,* which at 40 °C would lead to exceedingly high capacity factors at the column outlet.[15]

---

* Since the critical pressure of CO_2 is about 72 bar, this implies that a gaseous phase would leave the column. In our experience, low efficiencies are obtained and bad peak shapes are obtained when CO_2 enters the column as a supercritical fluid (temperatures 35 to 50 °C), but GC conditions prevail at the column outlet. This problem disappears when the temperature is far above the critical value. For example, helium ($T_C = -268$ °C) is frequently used in 'gas' chromatography with inlet pressures above the critical value ($P_C = 2.24$ atm).

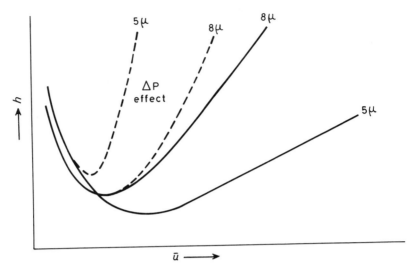

Figure 6 *Explanation of the observed efficiency vs. pressure drop curve of Figure 5a. The adverse effects of the column pressure drop are felt at much lower velocities for the 5 μm column because at the same velocity the pressure drop is higher* ($\Delta P \propto d_p^{-2}$), *whereas the optimum velocity is also higher* ($\bar{u}_{opt} \propto d_p^{-1}$).
(Reproduced by permission from *Chromatographia* in print.)[12]

Figure 5 is an illustration of the situation in which $\Delta P_{max} < \Delta P_{opt}$. This situation is explained in Figure 6. Before the plate height (h_r) can reach its minimum value, the adverse effects of the column pressure drop are felt. Columns for which $\Delta P_{max} < \Delta P_{opt}$ should not be used in SFC. At the given values for P and T, a column with larger particles may be used (see Figure 4a). The 5 μm column may be useful at higher mobile phase densities, where ΔP_{max} is higher.* Gere *et al.*[16] have measured *h* vs. *u* curves for columns packed with particles as small as 3 μm in SFC and have found that very high pressure drops (up to several hundred bar) could be used at high inlet pressures and low temperatures (high density mobile phases). Model calculations[8] reveal that at high mobile-phase densities the density of the mobile phase does not change significantly along the length of the column and that SFC appears to resemble LC with low viscosity liquids as the mobile phase.

It may be argued that variations in the observed capacity factor are merely of theoretical interest and that the observed peak width is of more practical value. Schwartz[17] has stated that the "inherently higher pressure drop in packed columns... should not be a deterrent to most applications". On the basis of our own work, we tend to conclude that this statement is correct, provided that columns are selected which allow operation at or above the

---

* At higher densities η increases, but this is more than compensated by a decrease in D_M, due to which the optimum linear velocity $\bar{u}$ decreases. Therefore, at higher densities ΔP_{opt} will decrease slightly, whereas ΔP_{max} will increase. In the extreme case of very high densities, *i.e.* liquid chromatography, very high pressure drops are of course allowed.

optimum pressure drop (*i.e.* $\Delta P_{max} > \Delta P_{opt}$). At low densities (where ΔP_{max} is low) or for complex separations (where the required number of plates, hence the required column length and hence ΔP_{opt} is high) capillary columns may be needed.

Instrumental Considerations

Flow Control

In principle, there are two degrees of freedom when operating an SFC column. With a given mobile phase, at a given temperature and pressure, two of the following three parameters may (within certain limits) be chosen by the chromatographer:

—the column inlet pressure,
—the column outlet pressure,
—the flow rate (or linear velocity) through the column.

It is of great importance to be able to use the two degrees of freedom to their full advantage. For SFC this is especially true. Because the diffusion coefficient (D_M) is a function of the density (see Figure 3), the optimum flow rate is not a constant for a given column, but a function of pressure and temperature. If only one of the two degrees of freedom is used, then much of the chromatography is likely to be performed at conditions far removed from the optimum values.[9]

By using a fixed restrictor behind the column, only the inlet pressure remains as a user-selectable variable. Under conditions of constant pressure, temperature, and composition, optimum flow rates may be achieved (in rather an inelegant manner) by changing the restrictor. However, in pro-grammed analysis the flow rate cannot be maintained at or near the optimum value.[9] Therefore, some kind of variable restrictor,[18] variable back-pressure regulator,[19] or flow controller is essential for optimum oper-ation of SFC systems.

It is quite a difficult proposition to regulate a small flow of a supercritical fluid accurately. However, it is not always impossible. Especially in situ-ations where (*a*) detection takes place under high pressure conditions (*i.e.* the restrictor is positioned after the detector, so that its 'dead' volume is not critical), and (*b*) the flow rates are relatively high (*e.g.* around 1 ml min⁻¹ of supercritical fluid), the flow rate can be accurately controlled. Currently, these two conditions are most likely to be met by packed-column SFC with UV or IR detection.

Figure 7 shows a chromatogram obtained with packed-column SFC with a constant (mass) flow, despite the use of a pressure programme. With the mass flow constant, the linear velocity through the column decreases proportionally with the mobile phase density. Assuming that to a first approximation over a limited range of densities the diffusion coefficient is inversely proportional to the density, this implies that $\bar{v}$ is kept approxi-mately constant throughout the analysis (see Equation 7a).

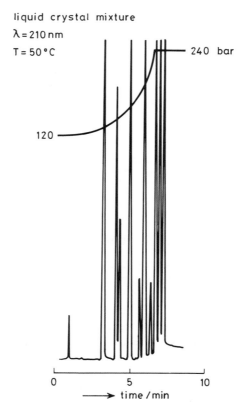

liquid crystal mixture
$\lambda = 210\,nm$
$T = 50\,°C$

Figure 7 *Chromatogram obtained with packed-column SFC using pressure programming. The mass flow of the eluent was kept constant with a two-stage mass flow controller. Sample: liquid crystal mixture. Column: 15 cm × 4.6 mm i.d., packed with 8 μm ODS silica*

Pumps

A constant pressure and pulseless operation are the main requirements for pumps used in SFC. Syringe pumps, with a single large piston, provide the best answer to the second demand. However, at high flow rates reciprocating piston pumps are more practical. For volumetric flow-rates below $1\,\mu l\,min^{-1}$ even syringe pumps may not provide a sufficiently accurate control, so that 'flow splitting' before the column becomes necessary. Some typical flow-rates for packed columns (at $\bar{v} = 10$) and for open-tubular columns (at $\bar{v} = 45$) are listed in Table 7 for the low-, medium- and high-density situations identified in reference 9.

Very low flow rates are seen to be required for open-tubular columns. The higher the mobile phase density, the lower the required flow rates. Fixed restrictors show the opposite behaviour, *i.e.* rapid increase in the flow rate upon increasing the column inlet pressure.[9] For all but the low density applications, the flow rates in packed-column SFC using conventional or narrow-bore columns are found to be in the typical operating range of HPLC pumps (*i.e.* between 0.1 and 10 ml min^{-1}).*

Low density applications of packed-column SFC can be performed very rapidly with high linear velocities. High volumetric flow rates are required for conventional columns, but not for narrow-bore columns. Because the maximum permissible pressure drop is smaller at low densities, such very rapid analyses will probably be limited to simple separations, requiring low numbers of theoretical plates.

Table 7 *Volumetric flow rates for a number of different SFC columns with carbon dioxide at low, medium, and high density as the mobile phase.[9] Low density:* $\rho = 0.28\,g\,cm^{-3}$, $D_M = 3.3 \times 10^{-4}\,cm^2\,s^{-1}$. *Intermediate density:* $\rho = 0.54\,g\,cm^{-3}$, $D_M = 0.78 \times 10^{-4}\,cm^2\,s^{-1}$. *High density:* $\rho = 0.79\,g\,cm^{-3}$, $D_M = 0.30 \times 10^{-4}\,cm^2\,s^{-1}$. $\bar{v} = 45$ *has been used for open-tubular columns,* $\bar{v} = 10$ *for packed columns. For conventional packed columns* $d_c = 4.6\,mm$, *narrow-bore packed columns* $d_c = 1\,mm$, *and for packed-capillary columns* $d_c = 0.3\,mm$. *The porosity* (ε) *was assumed to equal 0.6.*

Column type	Diameter[a]	Volumetric flow-rate (μl min^{-1})		
		Low density	Intermediate	High density
Open	8.5	0.59	0.14	0.054
	17	1.19	0.28	0.11
	50	3.5	0.83	0.32
	100	7.0	1.7	0.64
Packed	3	66,000	16,000	6,000
(Conventional)	5	39,000	9,300	3,600
	10	20,000	4,700	1,800
Packed	3	3,100	740	280
(Narrow-bore)	5	1,900	440	170
	10	930	220	85
Packed-capillary	3	280	66	25
	5	170	40	15
	10	84	20	7.6

[a] d_c for open columns, d_p for packed columns.

Injection and detection

The permissible injection volumes (V_{inj}) can be estimated as the (mobile phase) volume of one theoretical plate. Table 8 shows the resulting values for a number of columns. Routine valve injection can be performed down to volumes of about 0.1 μl. This kind of injection is therefore only possible with conventional packed columns. For smaller volumes rapid valve switching or splitting is required.

---

* SFC pumps are often operated at temperatures lower than that of the column, so that the volumetric flow rate delivered by the pump is lower than that passing through the column.

The operation of very narrow open columns with internal diameters of the order of 10 microns (as is required to compete with packed columns in terms of speed) appears to be very difficult. Both the injection and detection volumes need to be extremely small. This situation is similar to the one encountered in open-tubular LC.[11] It is therefore unlikely that narrow-bore capillary columns will become competitive with packed columns for simple, rapid separations by SFC. Rather, the potential use of such columns will be for the analysis of complex samples, requiring very high numbers of theoretical plates.* Narrow-bore capillary columns would allow practical analysis times and *if* FID could be made to work at the low (mass) flow rates involved, open-tubular SFC would have the sensitive, universal detector, which open-tubular LC has not.

Table 8 shows the permissible detection volumes for columns with $n = 20,000$ and solutes with $k' = 1$. Large flow cells are possible with conventional packed volumes, which, combined with the high loadability of the columns, is very attractive for UV and IR detection.[†] It is seen from the table that in terms of V_{inj} and V_{det} a packed-capillary column with 3μm particles behaves similarly to an open column with an internal diameter of 50 μm. This implies that the two columns can be used in the same instrument, provided that (*a*) a hundred-fold higher flow rate can be accommodated (Table 7), and that (*b*) the electronics of the detector are sufficiently fast to record the signal from the much faster packed column. This second point is illustrated in the last column of Table 8, which lists the required time constants for the detection system.

Programming Techniques

Programming techniques may be used in chromatography for the elution of samples that contain a wide range of different solutes (Chapter 6 in reference 10). In SFC this would apply if a sample contained both solutes which are eluted at low densities and solutes which require high mobile phase densities. A classic example is the elution of a homologous series of (*e.g.* polystyrene) oligomers. Programmed elution will allow all components of the series to be seen in a single chromatogram.

For a wide range sample such as a homologous series, programmed elution will reduce the required analysis time. Also, the sensitivity for different members of the series (which is related to peak width) will be much more constant throughout the chromatogram.

However, it should be emphasised that the resolution between any two solutes in the chromatogram can never be better in a programmed elution than in an elution with constant conditions and optimum capacity factors. Programmed elution must be used when the capacity factors cannot be

---

* The permissible detection volume increases proportional to $\sqrt{n}$.
† When the diameter of packed columns is increased, the permissible injection volume (V_{inj}) increases. Therefore, the concentration in the detector can be kept constant. According to Beer's law, the absorbance is proportional to the product of concentration and path length.

Table 8 *Permissible injection and detection volumes and permissible detection time constants for a number of different SFC columns. For open-tubular columns $v = 45$ and $h_r = 4.5$, for packed columns $v = 10$ and $h_r = 3$. V_{inj} was calculated as the volume of one theoretical plate ($\varepsilon = 0.6$ for packed columns). V_{det} was calculated as $0.5\sigma_v$ for $n = 20,000$ and $k' = 1$. τ_{det} was calculated as $0.2\sigma_t$ for $n = 20,000$, $k' = 1$ and $D_M = 2 \times 10^{-4}\,cm^2\,s^{-1}$. For Equations see Chapter 7 of reference 10.*

Column type	Diameter[a]	V_{inj} nl	V_{det} nl	τ_{det} ms
Open	8.5	0.002	0.3	20
	17	0.02	2.5	80
	50	0.4	63	700
	100	3.5	500	2800
Packed	3	90	13,000	8
(Conventional)	5	150	21,000	20
	10	300	42,000	85
Packed	3	4	600	8
(Narrow-bore)	5	7	1000	20
	10	14	2000	85
Packed-capillary	3	0.4	54	8
	5	0.6	90	20
	10	1.3	180	85

[a] d_c for open columns, d_p for packed columns.

brought in the optimum range for all solutes. The sacrifice in resolution by using programmed analysis is greater the higher the programming speed, *i.e.* the more rapidly the parameters (density or pressure in SFC) are changed with time. Fast programming[20] thus means that only a fraction of the potential separation power over a column is used.

The possible separation speed of a column will largely be determined by its hold-up time (t_0). It is not very sensible to squeeze all components in a complex mixture between t_0 and $1.1t_0$! It may be argued that when using a fixed restrictor, the actual value of t_0 may decrease during density programming,[9] but (as explained previously) this situation is, in fact, highly undesirable.

Thus the discussion about speed of analysis for different kinds of columns of different dimensions as presented in this chapter is equally valid for programmed and non-programmed analysis.

Packed-capillary Columns

The characteristics of packed-capillary columns are summarised in Table 9. In comparison with conventional packed columns, the main advantage is the higher specific permeability (Equation 13b).* As a consequence, a higher

* This effect has been observed by a number of workers, but has not yet been explained.

number of theoretical plates can be obtained with the same pressure drop over the column. Additional advantages are that interfacing with FID or MS detectors is made easier, and that the relatively low flow-rates required for packed-capillary columns (Table 7) allow the use of syringe pumps without the need to refill the syringe during the analysis.

Table 9 *Summary of the characteristics of packed-capillary columns.*

Compared to packed columns

Advantages
    Lower pressure drop
    Higher maximum number of theoretical plates
    More easily interfaced with FID or MS detectors
    Lower flow rates (syringe pumps)

Disadvantages
    Lower optimum velocity[13]
    Flow control more difficult
    Less sensitive with UV or IR detectors

Compared to open-tubular columns

Advantages
    (Much) shorter analysis times
    Higher sample loadability
    Possible to use UV or IR detection

Disadvantages
    Higher pressure drop
    Lower maximum number of theoretical plates
    Larger surface area (possible adsorption effects)

A disadvantage may be that the analysis times are slightly longer, because it has been observed that the optimum velocity is lower for packed-capillary columns.[13] Also, controlling the flow rate other than by a fixed restrictor is more difficult than with conventional packed columns. Finally, because the volumetric flow rates are much lower (Table 7), much smaller detector cell volumes will be required (Table 8), probably forcing the optical path length to be reduced. Because the concentrations of the solutes in the effluent stay the same (Table 2), the detector sensitivity will be reduced.

In comparison with open columns of current diameters (50 or 100 μm), packed-capillary columns offer much shorter analysis times (see Table 6). The much higher sample loadability of packed columns is favourable for concentration-sensitive detectors (*e.g.* UV or IR).

The pressure drop per unit number of plates, although lower than for conventional packed columns, is still much higher than for open columns. Therefore, the maximum number of theoretical plates that can be achieved with a packed-capillary column is less than with an open-capillary. However, if more than 100,000 theoretical plates can be obtained in a relatively short analysis time with packed-capillary columns (as is suggested by the data in Table 6), few applications of SFC will require open columns solely because of the required number of theoretical plates.

Selection of Columns

When selecting columns for application in SFC, the main part of the decision making should be based on the separation that is to be performed, *i.e.* (*a*) on the sample to be analysed and (*b*) on the type of analysis that is required.

The sample will determine the mobile phase required, the density required and whether or not programmed analysis is required to elute all components from the sample. Also, the number of plates required for a separation will depend on the complexity of the sample, or (for samples with a limited number of peaks) on the degree of similarity between the different sample components.

However, the type of analysis that is required is also relevant in this respect. For example, if a hydrocarbon group type analysis is required, the sample may be very complex, but it is not required to separate all the individual components.

If the required number of plates is very high, then capillary columns are required.* Also, if the required density is low, capillary columns may be attractive, because the pressure drop permitted over a packed column is limited. In that case relatively large particles must be used. Particles of 30 µm are comparable in speed of analysis to 50 µm open columns.

To select a packed column for SFC, the following path may be followed.

1. Determine the kind of stationary phase to be used, the mobile phase density required, and whether or not programmed elution is required. This needs to be done experimentally. Relatively short (*e.g.* 10 cm) columns packed with relatively large particles (*e.g.* 10 µm) may be used for this purpose. Different columns are required if different stationary phases are to be tested.

2. Select the appropriate pressure and temperature. The same density may be realised with different combinations of pressure and temperature. The higher the temperature, the higher is the required pressure to achieve a given density. Working at low temperatures may be attractive if a high density is required and very high pressures are to be avoided. Also, some solutes may not be thermally stable at elevated temperatures. (Relatively) high temperatures may sometimes be desirable to enhance the efficiency (*e.g.* the diffusion in the stationary phase), or in some cases to enhance the solubility of the solutes in the mobile phase.

3. Determine the minimum column outlet pressure at the selected temperature, for instance from the variation of the capacity factor with the pressure at different temperatures.[15]† The difference between the selected inlet pressure and the minimum outlet pressure determines the maximum permissible pressure drop over the column (EP_{max}).

---

* Currently, packed columns will easily allow up to 25,000 plates to be realised. For packed-capillary columns this number is expected to be considerably higher.
† At a given temperature there is a range over which k' varies more or less steeply with the pressure. Above this range it appears that SFC columns can be safely operated.

4. Select a column for which under the given conditions $\Delta P_{opt} < \Delta P_{max}$. This implies that if ΔP_{max} is large, columns with small particles can be used. If ΔP_{max} is small, larger particles and/or shorter columns are required. For a given column, the optimum pressure drop is not a strong function of the density because with increasing density $\bar{u}_{opt}$ decreases and η increases. The two effects largely cancel each other out.

Although general guidelines for the selection of columns can be formulated, many of the above recommendations are quite vague and require considerable insight and effort from the chromatographer. Much work is still required in order to acquire the general knowledge and experience with packed-column SFC that is required to formulate more practical schemes for the selection of columns.

4 Discussion and Conclusions

Columns for SFC

The comparison of packed and open columns for SFC as described in this chapter may be summarised by the following conclusions.

1. Based on the concept of reduced film thickness, open columns for SFC should have relatively thick films, both in comparison with GC columns and in comparison with columns currently used for SFC. The higher the density of the mobile phase in SFC, the higher is the desired film thickness in capillary columns. Stationary phases with better diffusion characteristics will allow higher film thicknesses.
2. To elute polar solutes from a packed column by SFC requires either the use of polar modifiers in the mobile phase, or (newly developed) stationary phases, which are more inert and more homogeneous. For the analysis of truly polar samples (*e.g.* body fluids) with any kind of SFC system, a breakthrough is required with respect to the use of polar supercritical fluids.
3. The speed of analysis of a capillary SFC column with a diameter d_c is comparable to that of a packed column with a particle size $d_p \simeq d_c/\sqrt{3}$. This implies that at present capillary SFC cannot compete with packed-column SFC in terms of speed of analysis.
4. For a given pressure drop over the column, contemporary capillary SFC (50 or 100 μm columns) will allow 100 to 500 times more theoretical plates than packed-column SFC with 5 or 10 μm particles. Packed-capillary columns may allow about 3 times as many plates as conventional packed columns.
5. The pressure drop over the column is a key issue in SFC. It has been found that the density of the mobile phase at the column exit should remain above a certain value. This means that for a given column at a given temperature and a given inlet pressure there is a maximum

permissible pressure drop. Columns packed with very small particles cannot be used for applications requiring low density mobile phases. Generally, the column pressure drop becomes the more important the closer the conditions are to the critical point.

6. Conventional packed columns may generate more than 20,000 plates within a few minutes for routine applications of SFC. Larger numbers of plates (possibly in excess of 100,000) may be obtained from packed-capillary columns within 1 hour. For even larger number of plates open columns will be necessary. Unless the column diameters can be greatly reduced, long analysis times are required for such high-efficiency separations.

General Application Areas for SFC

The discussion in the present chapter allows some conclusions on the general application areas of the two different SFC techniques. Table 10 reflects some speculations on the possible applications of capillary SFC. For packed-column SFC a similar list is shown in Table 11.

Table 10 *General application areas of capillary SFC.*

Open-tubular (capillary) SFC

Capillary SFC is most useful
    at relatively low densities
    for high-efficiency separations, *i.e.*
    for complex samples

Capillary SFC can best be used with mass flow-sensitive detectors:
    FID
    MS

Capillary SFC with universal detection (*e.g.* FID) can be used as an extension of capillary GC, *e.g.* for
    Fingerprinting of complex samples
    Samples which are not sufficiently volatile for GC
    Samples which are not sufficiently stable for GC

Capillary SFC may be used for identifying components in complex mixtures by SFC-MS

Capillary SFC is most useful at relatively low densities for a variety of reasons. Contemporary SFC columns are most compatible with low density mobile phases, because lower densities imply higher diffusion coefficients (Figure 1), so that columns of relatively large diameters may be used (Equation 16). The film thickness of contemporary capillary SFC columns is also more compatible with low density mobile phases (Table 5). In terms of speed of analysis packed columns are far superior (Table 6), but at low densities the permissible pressure drop over the column is limited.

Table 11 *General application areas of packed-column SFC.*

Packed-column SFC

Packed column SFC is most useful
   at relatively high densities
   for separations requiring low or moderate numbers of plates, *i.e.*
   for simple samples (containing a limited number of peaks).

Packed-column SFC can be used in combination with specific sample preparation and specific detectors to analyse a few components in a complex sample

Packed-column SFC provides the opportunity to use concentration-sensitive detectors (UV, IR) with good sensitivity.

Packed-column SFC may be also interfaced with FID and MS detectors

Packed-column SFC allows the separation of non-volatile samples
   on a routine basis
   more rapidly than HPLC
   with universal (FID) or almost universal detection (IR, UV at 190 nm)
   with specific detection, *e.g.* at specific wavelengths

Packed-column SFC may be used for identifying components in mixtures by SFC-MS, SFC-IR, and/or SFC-UV (diode array).

Because the maximum number of theoretical plates that can be obtained with an open column is much higher than with a packed column (Table 6), capillary SFC may be used for high-efficiency separations requiring (very) large numbers of theoretical plates. This suggests that complex samples, in which all components need to be separated may be analysed by capillary SFC.

Contemporary capillary SFC columns are very compatible with FID and MS detection. Advances in the design of pressure restrictors have made the use of these detectors on a routine basis possible. This allows capillary SFC to be applied to complex samples, which cannot easily be analysed by GC. However, to enable the analysis of complex polar samples, such as body fluids, much progress is still needed regarding the use of polar solvents or polar modifiers and the development of new stationary phases.

Packed column SFC is most useful in combination with high density mobile phases. In general, the closer SFC resembles LC, the more similar will be the column technology and the required instrumentation. As is the case with LC, SFC can be used for the routine analysis of samples containing a limited number of peaks. By applying specific sample preparation techniques or specific detection methods, complicated separation problems may be reduced to simple chromatograms.

In comparison with LC, packed-column SFC has the advantages that it is faster, that pressure (or density) is more easily changed or programmed than solvent composition, and that a variety of detectors can be readily applied. The problems associated with the interfacing of LC to sensitive, universal detectors (such as FID, MS) may create the most obvious opportunities for the successful application of SFC. Also, the solvents commonly used in HPLC obscure most of the IR spectral region, unlike some SFC solvents (Figure 1). SFC-IR can be used as a complement to SFC-MS, for example for the identification of isomers or for high molecular weight solutes.

Future of SFC

In the previous section a number of possible general application areas have been indicated for both capillary and packed-column SFC, indicating some directions in which SFC might develop. In only the last few years much progress has been made in SFC. In a paper published in 1987,[21] we identified a number of research goals for SFC. This list is reproduced as Table 12. Within the same year, progress has been recorded at scientific meetings and in journals on almost all the items listed.

Table 12 *Research goals for supercritical fluid chromatography.*[21]

General

Theory of supercritical solvents
Behaviour of supercritical mixtures
New (more polar) mobile phases
Non-corroding materials
Injection of solids

Packed-column SFC

Sufficiently inert stationary phases
Adequate flow or pressure control
Universal and sensitive detectors
Effects of column pressure drop
Effects of column and particle dimensions

Capillary SFC

Reproducible restrictors
Splitless injectors
Stable and efficient stationary phases
Narrow-bore columns ($d_c \leqslant 15 \, \mu m$)
Selective detectors

Applications

Establish unique applications (not covered by GC or LC)

Nevertheless, Table 12 is still a reasonable summary of the most relevant research areas in SFC. Added to the list could be the investigation of the potential of packed capillary columns and the combination ('hyphenation') of SFC with different spectrometric identification methods. Still, however, the last item on the list is the most relevant one for the immediate future of

SFC. The more applications can be demonstrated, which show SFC to be a unique and essential analytical method, the more rapidly will the technique mature along the lines described in this chapter.

References

1. C. L. Woolley, B. J. Tarbet, K. D. Bartle, K. E. Markides, J. S. Bradshaw, and M. L. Lee, in 'Proceedings of Eighth International Symposium on Capillary Chromatography' *ed.* P. Sandra, Huethig, Heidelberg, 1987, pp. 253–263.
2. H. H. Lauer, D. McManigill, and R. D. Board, *Anal. Chem.* 1983, **55**, 1370.
3. R. C. Wieboldt and D. A. Hanna, *Anal. Chem.* 1987, **59**, 1255.
4. S. M. Fields, R. C. Kong, M. L. Lee, and P. A. Peaden, *J. High Resolut. Chromatogr., Chromatogr. Comm.*, 1984, **7**, 423.
5. J. F. K. Huber, H. H. Lauer, and H. Poppe, *J. Chromatogr.*, 1975, **112**, 377.
6. Cs. Horváth and H.-J. Lin, *J. Chromatogr.*, 1976, **126**, 402, *ibid.*, 1978, **149**, 43.
7. P. A. Bristow, 'Liquid Chromatography in Practice', HETP, Wilmslow, 1976 p. 25.
8. P. J. Schoenmakers, P. E. Rothfusz, and F. C. C. J. G. Verhoeven, *J. Chromatogr.*, 1987, **395**, 91.
9. H. E. Schwartz, P. J. Barthel, S. E. Moring, and H. H. Lauer, *LC & GC*, 1987, **5**, 490.
10. P. J. Schoenmakers, 'The Optimization of Chromatographic Selectivity, A Guide to Method Development', Elsevier, Amsterdam, 1986.
11. H. P. M. van Vliet, 'Laser Induced Fluorescence Detection in Microbore and Open-Tubular Liquid Chromatography', Thesis, University of Amsterdam, 1986.
12. P. J. Schoenmakers and L. G. M. Uunk, '11th Symposium on Column Liquid Chromatography', Amsterdam, June 28–July 3, 1987. *Chromatographia*, accepted for publication.
13. M. Verzele and C. Dewaele, 'Proceedings of Eighth International Symposium on Capillary Chromatography', ed. P. Sandra, Huethig, Heidelberg, 1987, pp. 1075–1093.
14. P. J. Schoenmakers and F. C. C. J. G. Verhoeven, *J. Chromatogr.*, 1986, **352**, 315.
15. P. J. Schoenmakers, *J. Chromatogr.*, 1984, **315**, 1.
16. D. R. Gere, R. Board, and D. McManigill, *Anal. Chem.* 1982, **54**, 736.
17. H. E. Schwartz, *LC & GC*, 1987, **5**, 14.
18. T. Greibrokk *et al.*, *J. Chromatogr.*, in press.
19. D. R. Gere, Personal communication, 1984.
20. R. D. Smith, H. T. Kalinoski, H. R. Udseth, and B. W. Wright, *Anal. Chem.*, 1984, **56**, 2476.
21. P. J. Schoenmakers and F. C. C. J. G. Verhoeven, *Trends Anal. Chem.*, 1987, **6**, 10.

CHAPTER 5

HT-CGC and CSFC for the Analysis of Relatively High Molecular Weight Compounds

P. SANDRA, F. DAVID, F. MUNARI, G. MAPELLI
and S. TRESTIANU

1 Introduction

Recent developments in the analysis of molecular weight compounds ranging from 500 to 1500 daltons diverge into two mainstreams, high temperature capillary gas chromatography (HT-CGC) on the one hand, and capillary supercritical fluid chromatography (CSFC) on the other hand. Capillary gas chromatography generally is believed to be restricted to the analysis of 'volatile' compounds (MW up to 500), and less applicable to the analysis of so-called 'heavy' components (MW 500–1500).

New developments in capillary GC, however, especially the introduction of thermostable stationary phases and non-discriminative injection devices, have made the definitions of 'volatile' and 'heavy or less-volatile' very flexible. The applicability of capillary GC has been extended to MWs up to 1500 daltons. A prerequisite in HT-CGC, of course, is that the solutes to be analysed are thermally stable. The thermal stability of organic compounds in a completely inert system, *i.e.* fused silica, inert gases, highly purified stationary phases, is however much higher than currently accepted. The possibilities of HT-CGC (up to 420 °C–430 °C) have been illustrated by the analyses of hydrocarbons and crude oils,[1-5] of lipids,[5-12] of sterols and sterol esters,[5,9] of waxes,[13] of sugars,[9] of glycerols,[9] of oligomers,[5] of sterol glucosides,[14] of polymer additives,[15,16] of emulsifiers,[17] *etc.*

Parallel to the developments in HT-CGC, considerable progress has been made in supercritical fluid chromatography (SFC). The introduction of high quality syringe pumps, narrow-bore capillary columns with immobilised stationary phases, density and pressure programming, *etc.*, have made capillary SFC (CSFC) accessible to everyday chromatography. The interest in CSFC, therefore, continuously increases. Due to the high solvating power of supercritical fluids, high molecular weight (HMW) compounds can be eluted at much lower temperatures than in HT-CGC. CSFC has been applied to the analysis of HMW hydrocarbons,[18,19] of glycerides,[20-23] of dyes,[24] of polysaccharides,[25] of waxes,[25] and oligomers,[27,28] *etc.* In the recent literature, HT-CGC and CSFC are considered competitive to each other. However, a comparison of the features of both techniques, especially in quantitation, has never been made.

HT-CGC and CSFC have been compared for a number of applications.[29-31] The compounds investigated are listed in Table 1.

Table 1 *HT-CGC – CSFC: compounds considered*

— HYDROCARBONS
— LIPIDS
— PHOSPHOLIPIDS*
— ETHOXYLATED ALCOHOLS
— WAXES
— EMULSIFIERS
— AMINES
— QUATERNARY AMMONIUM SALTS*
— STEROIDS
— STEROLS AND ESTERS

---

* ionic species

In this contribution, some of the data are presented, and the results of HT-CGC and CSFC are compared, in terms of resolution, speed of analysis, applicability range, and quantitation.

2 Equipment for CSFC and HT-CGC

Analyses were performed on a SFC 3000 series chromatograph from Carlo Erba. This instrument has multipurpose chromatography capabilities, *i.e.* SFC with micropacked and capillary columns and HT-CGC (Figure 1).

The SFC part consists of an SFC 300 pulse-free syringe pump, an independent valve (split or direct) injection system, column oven, interchangeable optimised GC detectors, an ICU 600 control unit, and computer control-data handling. The mobile phase solvation strength is tunable through multiramp density or pressure programming, and multiramp positive or negative temperature programming.

The HT-CGC part is equipped with a cold on-column injector with an elongated secondary cooling tube enabling high oven temperature (H.O.T.) cold on-column injections.[12]

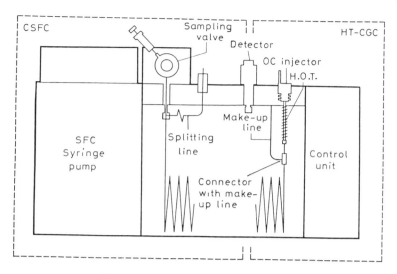

Figure 1 *HT-CGC–SFC instrument*

Table 2 *Analytical conditions*

	HT-CGC	CSFC
Injection	On column (H.O.T.)	Valve split injection
	Autosampler	Autosampler
Column	Glass	
	Fused silica – PI	Fused silica – PI
	Fused silica – Al	
	5–25 m	5–20 m
	0.2–0.32 mm i.d.	0.05–0.1 mm i.d.
Precolumn	20–30 cm	
	0.53 mm i.d.	
Phases	OV-1	OV-1
	OV-73	OV-73
	OV-17	OV-225
Oven	HT-MEGA	HT-MEGA
Detector	FID	FID
Interface		Open restrictor
		fused silica
		5–10 μm i.d.
Carrier	He, H_2	
	(CP-CF 515)	CO_2

Both CSFC and HT-CGC analyses can be performed automatically *via* the AS 550 autosampler. For automated HT-CGC, the analytical column is connected to a 20–30 cm × 0.53 mm i.d. precolumn *via* a butt connector equipped with auxiliary carrier make-up gas. The GC capillary columns were coated with 0.1 μm dimethyl silicone (OV-1 or SE-54) or 0.1 μm phenyl-methyl silicone (similar to OV-17).[32] The maximal allowable operating temperatures are, respectively, 420 °C for dimethyl silicone and 370 °C for phenylmethyl silicone. Polyimide coated fused silica columns were used for temperatures up to 390 °C; borosilicate- or aluminium-clad columns were used for temperatures exceeding 380 °C–390 °C. CSFC was performed on fused silica open-tubular (FSOT) capillary columns (i.d. 50 or 100 μm) coated with dimethyl silicone or cyanopropylmethyl silicone (OV-225). A straight restrictor (5 or 10 μm i.d.) was connected to the column outlet. Flame ionization detection (FID) at 350 °C to 400 °C was used for both HT-CGC and CSFC. The analytical conditions are summarised in Table 2.

3 Discussion of Separations by CSFC and HT-CGC

Hydrocarbons

Both HT-CGC[1,3] and CSFC[18,19] have been put forward for the calculation of the true boiling point distribution of heavy petroleum products (simulated distillation). In this type of analysis, high resolution is not required. The most important prerequisite the technique should fulfil, is complete and quantitative elution.

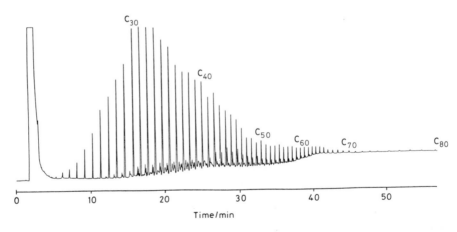

Figure 2 *HT-CGC–wax analysis. Column: 25 m × 0.32 mm i.d., SE-52; temperature: 178 °C (1 min) to 410 °C (10 min) at 7 °C min⁻¹; carrier gas: helium, 2 ml min⁻¹*
(Reproduced by permission from *LC & GC*, submitted for publication)[29]

Hydrocarbons up to C_{130} elute quantitatively in HT-CGC on short capillary columns coated with a thin film of dimethyl silicone.[1,3] This technique is nowadays well-established as simulated distillation (SIMDIS). It has been claimed that the temperatures needed (up to 430 °C) to elute the high carbon numbers, cause thermal decomposition. With high oven temperature cold on-column injection onto properly deactivated capillary columns, decomposition and also sample discrimination of HMW petroleum residues does not occur. On the other hand, in CSFC, sample discrimination according to the carbon number is difficult to avoid with valve split injection. Figure 2 shows the high resolution HT-CGC analysis of a paraffin wax containing hydrocarbons from C_{20} to C_{80}. In automated analysis ($n = 4$), the RSD(%) is 0.1 for retention times and 3.5 for absolute peak areas (Table 3).

Table 3 *Peak area and retention time reproducibility in automated HT-CGC analysis of paraffin waxes. For chromatographic conditions, see Figure 2.*[29]

	C_{30}	C_{35}	C_{40}	C_{45}	C_{50}
Area (counts)	160 917	152 534	87 030	35 680	14 002
	158 497	150 140	86 040	36 456	14 619
	159 878	151 574	85 396	37 388	14 890
	155 732	148 235	86 523	38 079	14 043
Average (counts)	159 694	150 620	86 247	96 899	14 561
SD (counts)	1 000	1 870	696	1 050	517
RSD (%)	0.62	1.2	0.80	2.8	3.5
Retention time (min)	17.83	21.99	25.62	30.01	32.79
	17.80	21.97	25.60	30.00	32.76
	17.79	21.96	25.59	29.99	32.75
	17.79	21.95	25.58	29.98	32.74
Average (min)	17.80	21.96	25.60	29.99	32.76
SD (min)	0.019	0.017	0.018	0.012	0.023
RSD (%)	0.10	0.07	0.07	0.04	0.07

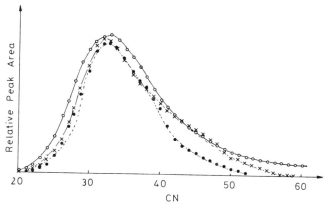

Figure 3 *MW distribution of wax sample* ○: *HT-CGC;* ●: *CSFC; injection valve at 22 °C; X: CSFC; injection valve at 60 °C*

The same analysis was carried out with CSFC. The MW distribution (relative peak area as a function of carbon number) calculated from CSFC data, with the injection valve at ambient temperature (22 °C) and at 60 °C, and from the HT-CGC data is shown in Figure 3.

From the curves it is obvious that valve split injection leads to important carbon number discrimination. With the valve at ambient temperature, the last compound that can be detected is C_{52}. With the valve at 60 °C, hydrocarbons elute up to C_{59}. The HT-CGC curve, on the other hand, has a much flatter slope at high carbon numbers and hydrocarbons up to C_{80} are recorded. The importance of the valve temperature was discussed by Schwartz *et al.*[19] Sample discrimination in CSFC is mainly due to an incomplete transfer (condensation!) of the sample components from the sample loop into the column and the higher the valve temperature, the better is the transfer (volatility contribution!). An illustration of sample discrimination by valve injection is given by the analysis of Polywax 500 (Figure 4). The analysis was carried out on a short capillary column, *i.e.* in the low resolution mode.

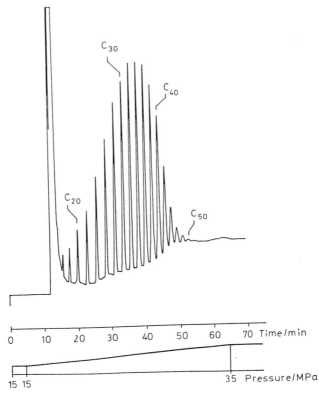

Figure 4 *CSFC–Polywax* 500. *Column:* 5 m × 0.1 mm *i.d., OV-73; temperature:* 100 °C; *fluid:* CO_2; *linear pressure programme*: 15 MPa *to* 35 MPa *at* 0.33 MPa min^{-1}
(Reproduced by permission from *LC & GC*, submitted for publication)[29]

With the valve at room temperature, the maximum in the MW distribution is found at C_{34}. Peak areas rapidly decrease after C_{40}, resulting in a non-Gaussian distribution profile. The curve obtained with HT-CGC (see reference 2) shows a Gaussian distribution with maximum at C_{40} and last peak C_{70}. To obtain the same distribution with CSFC, the injector valve (and the syringe!) have to be heated above 100 °C. For simulated distillation, the valve and syringe temperature must be much higher. This is not a very practical approach. New developments in sample introduction in CSFC, *i.e.* direct injection, seem to overcome this problem.[33,34]

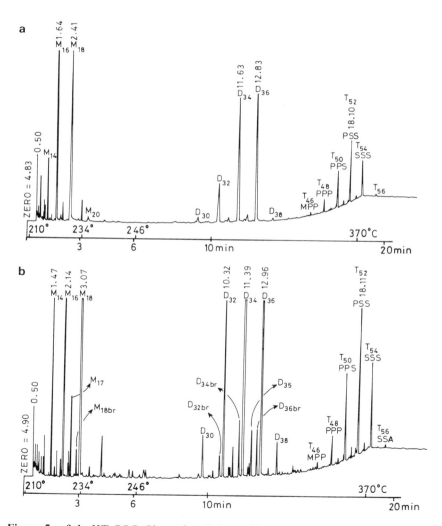

Figure 5 a & b *HT-CGC–Glycerides. Column:* 25 m × 0.25 mm i.d., OV-1; *temperature:* 210 °C to 370 °C; *carrier gas: hydrogen,* 1 bar
(Reproduced by permission from 'Sample Introduction in Capillary Gas Chromatography', Vol. 1 Huethig, 1985, p. 159)[9]

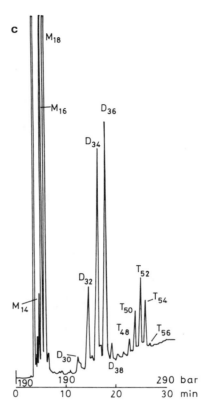

Figure 5 c *CSFC–Glycerides.   Column:*   10 m × 0.1 mm *i.d.,   SE-54;   temperature:*
170 °C; fluid: CO_2; *linear pressure programme* 19 MPa *to* 29 MPa *at*
0.33 MPa min^{-1}
(Reproduced by permission from *LC & GC*, submitted for publication)[29]

Lipids

The qualitative and quantitative elucidation of glycerides is an important analysis in different fields, *i.e.* characterisation of natural products, of food products, of emulsifiers, in lipid metabolism studies, and bacterial identification, *etc.* A variety of techniques is routinely applied including thin layer chromatography (TLC), high performance liquid chromatography (HPLC), and HT-CGC. In recent years, CSFC has gained considerable interest for the characterisation of glycerides.[20,21,23]

Of these techniques, HT-CGC provides the highest resolution in the shortest analysis time. The analyses of a standard mixture of mono-, di- and triglycerides by HT-CGC (unsilylated [**a**] and silylated [**b**]) and CSFC (**c**) are shown in Figure 5.

The HT-CGC analyses were carried out with optimised carrier gas flow conditions, whereas the CSFC trace was recorded at CO_2 velocities far above the optimal value. Under optimal flow conditions, both columns theoretically offer the same plate number (25 m × 0.25 mm i.d. *vs.* 10 m × 0.1 mm i.d., $n = L/h$, $h = d_c$, $n = 100,000$ plates). At h_{min}, the optimal mobile phase velocity ($\bar{u}_{opt}$) in CGC with hydrogen is in the range 30–50 cm sec^{-1} whereas in CSFC with CO_2, $\bar{u}_{opt}$ is 0.1 to 0.3 cm sec^{-1}. The $h\bar{u}$ van Deemter plot for the column used is shown in Figure 6. Working close to the optimal $\bar{u}$ value (a) would result in analysis times of several hours. A mobile phase velocity of 4 to 5 cm sec^{-1}(b) was selected to give analysis times in the order of 30 min; unfortunately the resulting plate number is only 5,000 to 6,250. Further developments should be directed to the optimisation of very small internal diameter columns. Figure 7 shows theoretical van Deemter plots for capillary columns with different diameters.[35] A column of 1 m × 10 μm i.d. should offer 100,000 plates at a mobile phase velocity of 1 cm sec^{-1}, which guarantees short analysis times. At present, technological difficulties are encountered in the application of such columns, but developments are moving very rapidly.

HT-CGC and CSFC, applying the possibilities available nowadays, were compared for the analysis of the triglycerides of milk chocolate (Figure 8). HT-CGC is routinely applied in the chocolate industry because of the higher efficiency and the much shorter analysis time (only 3 min). The analysis allows the determination of the quantity of milk fat in chocolate (asterisk-marked peaks) as well as the presence of cocoa butter equivalents.

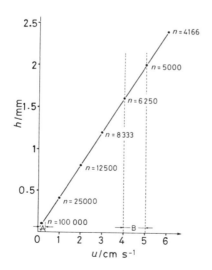

Figure 6 *Van Deemter plot for* 10 m × 0.1 mm *i.d. column. Stationary phase: SE-54, mobile phase:* CO_2

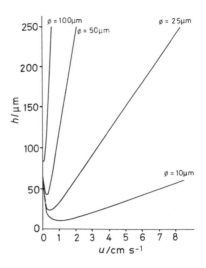

Figure 7 *Van Deemter plots for capillary columns with different diameters*

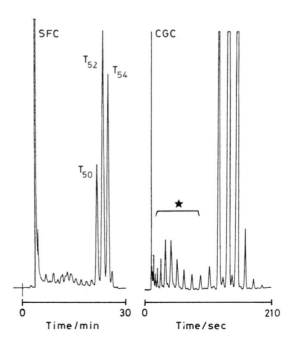

Figure 8 *CSFC and HT-CGC–Milk chocolate triglycerides. CSFC: column:*
10 m × 0.1 mm *i.d., SE-54; temperature:* 170 °C; *fluid:* CO_2, *linear pressure*
programme 19 MPa *to* 29 MPa; *quantitation:* T_{50} : 18.77%, T_{52} : 45.63%,
T_{54} : 35.50%. *HT-CGC: column:* 6 m × 0.25 mm *i.d., OV-1; temperature*
290 °C *to* 350 °C *at* 20 °C min^{-1}; *carrier gas: hydrogen,* 1 bar; *quantitation:*
T_{50} : 18.01%, T_{52} : 46.37%, T_{54} : 35.62%

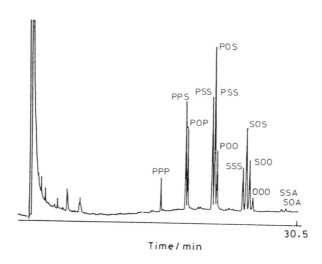

Figure 9 *HT-CGC automated analysis of lipids. Column:* 25 m × 0.25 mm *i.d., OV-17; temperature:* 280 °C *to* 350 °C *at* 3 °C min⁻¹; *carrier gas: hydrogen:* 0.7 bar (Reproduced by permission from *J. High Resolut. Chromatogr., Chromatogr. Commun.,* 1987, **10**, 263)[12]

It is often claimed that HT-CGC yields erratic quantitative results because of the decomposition of triglycerides. Problems can be encountered with oils containing large amounts of highly unsaturated triglycerides such as trilinolenin (LnLnLn), which tend to polymerise, not to decompose. For quantitation of such lipids, calibration is necessary. Most of the oils and fats, however, can be analysed perfectly well by HT-CGC, as proven by the similarity in quantitation of the main triglycerides in the HT-CGC and CSFC trace of milk-chocolate (Figure 8). In HT-CGC on apolar columns of the dimethyl silicone type, triglycerides are separated according to the carbon number (CN). On polarisable phenylmethyl silicone phases,[32] besides a CN-number separation, lipids are also separated according to the different combinations of saturated and unsaturated fatty acids in the triglycerides. Figure 9 shows the automated analysis of transesterified cocoa butter. The repeatability for areas and retention times for five consecutive runs is given in Table 4.

At present, the same resolution cannot be obtained by CSFC, although selective interaction sites, *i.e.* phenyl and cyanopropyl groups, can be introduced into the stationary phase yielding separation according to the degree of unsaturation. Figure 10 shows the separation on OV-225 of tristearin (SSS), triolein (OOO), trilinolein (LLL) and trilinolenin (LnLnLn).

Table 4 *Peak area and retention time reproducibility in automated HT-CGC analysis of triglycerides. For chromatographic conditions, see Figure* 9.[12]

Sample		A081	A082	A083	A084	A085	Mean	
PPP	RT	8.87	8.87	8.88	8.87	8.87	8.87	
	% dev.						0.05	
	Area	164.51	168.94	170.66	175.12	164.16	168.68	
	% dev.						2.71	
PPS	RT	10.88	10.89	10.91	10.89	10.89	10.89	
	% dev.						0.10	
	Area	632.42	637.72	646.03	661.02	642.53	643.94	
	% dev.						1.68	
POP	RT	11.08	11.09	11.11	11.09	11.09	11.09	
	% dev.						0.10	
	Area	503.85	492.63	493.76	507.84	494.97	498.61	
	% dev.						1.36	
PSS	RT	13.43	13.44	13.45	13.44	13.44	13.44	
	% dev.						0.05	
	Area	801.80	808.37	810.96	801.21	805.05	805.48	
	% dev.						0.52	
POS	RT	13.68	13.69	13.71	13.69	13.69	13.69	
	% dev.						0.08	
	Area	1189.36	1194.34	1205.45	1189.70	1197.93	1195.36	
	% dev.						0.56	
POO	RT	13.92	13.93	13.96	13.93	13.93	13.93	
	% dev.						0.11	
	Area	427.87	419.55	428.99	369.86*	410.65	421.76	(411.38)
	% dev.						2.02	(5.92)
SSS	RT	16.48	16.49	16.53	16.51	16.51	16.50	
	% dev.						0.12	
	Area	344.83	350.30	334.34	345.53	348.85	344.77	
	% dev.						1.81	
SOS	RT	16.83	16.84	16.87	16.83	16.83	16.84	
	% dev.						0.10	
	Area	745.56	750.55	747.89	762.66	751.37	751.61	
	% dev.						0.88	
SOO	RT	17.13	17.15	17.17	17.15	17.15	17.15	
	% dev.						0.08	
	Area	452.90	451.95	441.89	459.14	455.14	452.20	
	% dev.						1.41	
OOO	RT	17.47	17.48	17.51	17.49	17.49	17.49	
	% dev.						0.08	
	Area	127.88	116.66	111.02	118.92	120.37	118.97	
	% dev.						5.14	

Mean % deviations for all retention times: 0.09%.
Mean % deviations for all peak areas: 1.81%.
*: abnormal value, not taken into account.

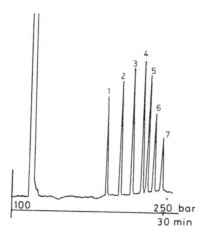

Figure 10 *CSFC–Triglycerides. Column:* 10 m × 0.1 mm *i.d., OV-225; temperature:* 150 °C; *fluid:* CO_2; *linear pressure programme:* 10 *to* 25 MPa *at* 0.5 MPa-min^{-1}. *Compounds:* 1. *trilaurin,* 2. *trimyristin,* 3. *tripalmitin,* 4. *tristearin,* 5. *triolein,* 6. *trilinolein,* 7. *trilinolenin*

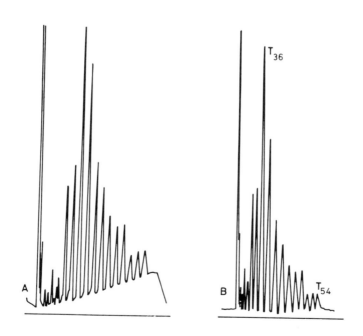

Figure 11 *CSFC–Palm kernel oil triglycerides. Column: A* 25 m × 0.1 mm *i.d., OV-73, B* 5 m × 0.1 mm *i.d., OV-73; temperature: A* 150 °C *isothermal, B* 130 °C *to* 50 °C *at* 3 °C min^{-1}; *fluid:* CO_2 *A* 19 MPa *to* 29 MPa *at* 0.33 MPa min^{-1}, *B* 20 MPa *isobaric*
(Reproduced by permission from *LC & GC,* submitted for publication)[29]

The separation number between SSS and OOO, however, is only 1. SOS and SOO cannot be located between SSS and OOO, as is the case in the HT-CGC analysis (Figure 9), where the separation number between SSS and OOO is 3. Another advantage of HT-CGC compared to CSFC is that the solutes, in first instance, are separated according to their vapour pressure (carbon number separation) and then, in their carbon number, separated according to their degree of unsaturation. In the case of Figure 10, a saturated triglyceride with carbon number 56 would overlap the unsaturated triglycerides with carbon number 54.

CSFC is ideally suited to analyse triglycerides according to their CN-number. Figure 11 shows the analysis of palm kernel oil in a density programmed run. To increase the density of the mobile phase, either the pressure can be increased (Figure 11A) or the temperature decreased (Figure 11B). The quantitative aspect of a CSFC analysis, is illustrated in Table 5, which also shows the reproducibility of retention data: both are excellent.

Table 5 *Reproducibility of retention data and quantitation in CSFC analysis of triglycerides.*[22]

	% T_{36}	% T_{42}	% T_{48}	% T_{54}	t_{R36}	t_{R42}	t_{R48}	t_{R54}
1	20.76	24.89	26.03	28.32	16.41	20.84	24.44	27.55
2	20.71	24.55	26.41	28.32	16.39	20.83	24.44	27.54
3	20.90	24.84	26.36	27.89	16.41	20.86	24.44	27.54
4	20.63	24.73	26.30	28.34	15.45	20.86	24.42	27.53
X	20.75	24.75	26.27	28.22	16.41	20.85	24.43	27.54
σ	0.1134	0.1506	0.1694	0.2185	0.0251	0.015	0.010	0.008
σ_{rel} %	0.547	0.609	0.645	0.774	0.153	0.072	0.041	0.030

These data were measured on a 10 m × 100 μm SE-54 column. Column temperature: 150 °C isotherm; fluid CO2. Pressure programme: 190 bar isobaric for 10 minutes, then programmed from 190 bar to 290 bar at 5 bar min[-1]. *Compounds: trilaurin, trimyristin, tripalmitin, tristearin*
(Reproduced by permission from *J. High Resolut. Chromatogr., Chromatogr. Commun.,* 1986, **9,** 189)[22]

An important aspect in the analysis of oils and fats is the determination of phospholipids (phosphatidylcholine, phosphatidylserine, phosphatidyl-inositol and phosphatidylethanolamine). Phospholipids cannot be analysed as such either by HT-CGC or by CSFC using CO_2. Derivatised phospholipids, on the other hand, can be analysed with CSFC as illustrated by the analysis of a phosphatidylethanolamine (Figure 12).

HT-CGC was not successful in this respect due to breakdown of the C–O–P bond under high temperatures. Full details on the derivatisation of phospholipids will be published elsewhere.

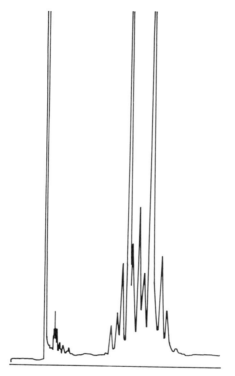

Figure 12 *CSFC–Phosphatidylethanolamine. Column:* 20 m × 0.1 mm *i.d., OV-73;*
temperature: 130 °C; *fluid:* CO_2, *linear pressure programme:* 18 MPa
(5 min) *to* 28 MPa *at* 0.35 MPa min^{-1}

Ethoxylated Compounds

The separation of the surfactant Triton X-100 (MW ranging from 250 to
1100) is often used to demonstrate the possibilities of CSFC (Figure 13a).
Figure 13b shows the chromatograms of Triton X-100 with HT-CGC. The
derivatisation into the TMS-derivatives is not a prerequisite for HT-CGC.[29]
Both chromatograms show a Gaussian distribution at $(-CH_2-CH_2-O-)_n$
($n = 20$). The relative standard deviations were calculated for retention times
and raw peak areas. The results are listed in Tables 6 and 7.

In automated HT-CGC, the RSD(%) for retention times is better than
0.15%, and for absolute peak areas better than 1.5%. In automated CSFC,
the values are respectively 0.15% and 4%. Concerning the resolution, it is
obvious from Figure 13 that a more detailed picture is obtained with HT-
CGC. The small peaks marked with an asterisk constitute a second Gaussian
profile, whereas this profile cannot be measured in the CSFC chromatogram.
The resolution decreases as the density increases; the higher the density, the
lower the efficiency (LC-like) and the lower the density, the higher the
efficiency (GC-like).

Table 6 *Peak area and retention time reproducibility in automated HT-CGC analysis of Triton X-100.*[29]

H.O.T. on column – AS 550 autosampler 6 m × 0.32 mm *i.d., SE-52 Oven temperature:* 160 °C (1 min) *to* 395 °C *at* 8 °C min^{-1} *(MEGA-HT 5350).Carrier gas: helium,* 2 ml min^{-1}

	Peak 1	Peak 2	Peak 3		Peak 1	Peak 2	Peak 3
	185 395	310 297	163 842		728.1	1 129.0	1 370.1
	182 635	309 225	165 083	Retention	727.9	1 129.6	1 370.5
Area	180 032	304 583	159 784	time	727.5	1 129.3	1 370.4
(counts)	180 282	304 579	163 191	(sec)	726.4	1 127.8	1 369.3
	185 886	315 802	165 278		726.4	1 127.7	1 369.4
Average							
(counts)	182 846	308 897	163 435	Average (sec)	727.2	1 128.6	1 369.9
SD (counts)	2 751	4 663	2 217	SD (sec)	0.815	0.877	0.563
RSD (%)	1.5	1.5	1.35	RSD (%)	0.11	0.07	0.04

(Reproduced by permission from *LC & GC*, submitted for publication)[29]

Table 7 *Peak area and retention time reproducibility in automated CSFC analysis of Triton X-100.*[29]

Column: 10 m × 0.1 mm *i.d., SE-52. Temperature:* 140 °C, *Fluid:* CO_2, *linear density programme* 0.22 g ml^{-1} (10 min) *to* 0.55 g ml^{-1} *at* 0.007 g ml^{-1}

	Peak area (counts)				Retention time (min)		
	Peak 1	Peak 2	Peak 3		Peak 1	Peak 2	Peak 3
	1 233 802	413 827	252 344	1	23.55	33.35	42.77
	2 228 027	396 081	240 911	2	23.55	33.35	42.77
	3 223 082	390 032	237 446	3	23.55	33.35	42.77
	4 223 782	393 184	236 809	4	23.55	33.35	42.77
	5 225 022	396 929	236 917	5	23.55	33.35	42.77
Run	6 229 213	404 829	242 670	Run 6	23.55	33.35	42.77
	7 224 767	397 498	236 401	7	23.55	33.35	42.77
	8 227 182	402 626	242 041	8	23.55	33.35	42.72
	9 237 945	422 477	249 062	9	23.55	33.37	42.77
	10 222 934	398 159	236 669	10	23.55	33.35	42.77
	11 228 029	405 559	243 277	11	23.55	33.35	42.77
	12 215 571	382 911	230 262	12	23.57	33.37	42.77
	13 240 396	417 188	241 374	13	23.59	33.42	42.82
	14 237 651	421 780	250 514	14	23.57	33.39	42.80
	15 221 707	397 400	237 263	15	23.55	33.35	42.77
	16 221 206	395 449	235 698	16	23.52	33.35	42.77
	17 247 242	458 207	264 357	17	23.59	33.39	42.80
	18 230 434	406 342	239 989	18	26.61	33.42	42.94
	19 228 535	407 313	236 663	19	23.59	33.39	42.82
	20 236 933	399 777	218 446	20	23.59	33.39	42.92
Average				Average			
(counts)	229 173	405 378	240 455	(counts)	23.56	33.37	42.79
SD (counts)	7 737	16 108	9 193	SD (min)	0.02	0.02	0.05
RSD (%)	3.4	4.0	3.8	RSD (%)	0.09	0.07	0.12

(Reproduced by permission from *LC & GC*, submitted for publication)[29]

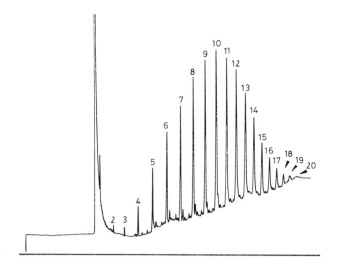

Figure 13a *CSFC–Triton X-100. Column:* 25 m × 0.1 mm *i.d., OV-73; temperature:* 100 °C; *fluid:* CO_2, *linear pressure programme:* 15 MPa (25 min) *to* 35 MPa *at* 0.2 MPa min^{-1}
(Reproduced by permission from *LC & GC*, submitted for publication)[29]

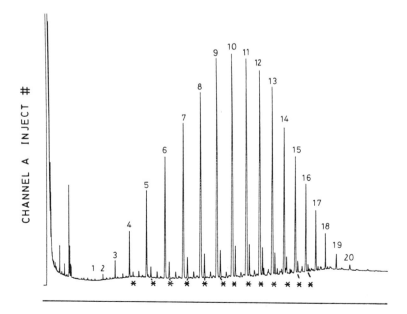

Figure 13b *HT-CGC–Triton X-100. Column:* 10 m × 0.32 mm *i.d., SE-52; temperature:* 65 °C (1 min) *ballistically to* 200 °C, *then to* 390 °C *at* 8 °C min^{-1}; *carrier gas: helium,* 2 ml min^{-1}
(Reproduced by permission from *LC & GC* submitted for publication)[29]

In conclusion, as long as the MW of ethoxylated compounds doesn't exceed 1400 daltons, HT-CGC is the method of choice. For higher MW mixtures, CSFC has to be preferred.

Waxes

Although waxes such as beeswax (Figure 14), can be analysed both by HT-CGC and CSFC,[29] we observed an important feature of CSFC during this comparison.

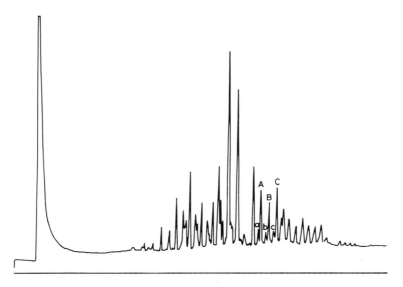

Figure 14 *CSFC–Beeswax. Column:* 10 m × 0.05 mm *i.d., OV-1; temperature:* 100 °C; *fluid:* CO_2, *linear pressure programme* 10 MPa (7.5 min) *to* 30 MPa *at* 0.4 MPa min^{-1}
(Reproduced by permission from *LC & GC*, submitted for publication)[29]

The peak pairs a-A, b-B, c-C of Figure 14 are better separated with CSFC and show a reversed elution order compared to HT-CGC. The main reason is supposed to be the lower analysis temperature in CSFC resulting in a higher selectivity of the stationary phase compared to HT-CGC. Selectivity of the mobile phase CO_2 in CSFC may not be excluded. On the other hand, selective phases such as cyanopropylphenylmethyl silicone, biscyanopropyl-(60%)vinylmethyl silicone, polyethylene glycol, *etc.*, which can readily be immobilised for SFC, cannot be used at temperatures above 250 °C–280 °C. Therefore they have no applicability in HT-CGC.[29]

Quaternary Ammonium Salts and Long Chain Amines

To our astonishment, a recent CSFC advertisement contained a CSFC chromatogram of quaternary ammonium salts obtained with CO_2 as the mobile phase. According to our experience, these ionic species could neither be analysed with CSFC-CO_2 nor with HT-CGC applying cold on-column injection. HT-CGC can be used for this application but only in combination with a vaporizing sample introduction system operated at high temperatures ($> 350\,°C$). The quaternary ammonium salts are demethylated in the injection port and the resulting tertiary amines elute nicely (Figure 15). For CSFC, ammonia seems to be suitable as a supercritical fluid mobile phase to elute quaternary ammonium salts.[36] It is questionable, however, whether the salts are chromatographed as such or whether the compounds are immediately desalted by the alkaline mobile phase.

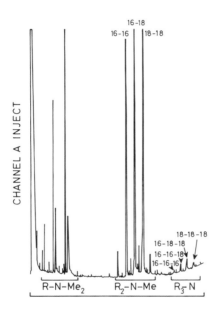

Figure 15 *HT-CGC–Quaternary ammonium salts. Column:* 5 m × 0.1 mm *i.d., OV-1, temperature:* 100 °C *to* 350 °C *at* 10 °C min^{-1}; *carrier gas: hydrogen,* 2 bar; *split-injection at* 350 °C
(Reproduced by permission from *J. High Resolut. Chromatogr., Chromatogr. Commun.*, to be published)[30]

Quaternary ammonium salts can easily be converted into the amines by off-line desalting with phosphoric acid at elevated temperatures. The amines can be analysed with CSFC using CO_2, as illustrated in Figure 16.

The analysis of tertiary amines does not create any problem in CSFC; this, however, is not the case for other amines such as primary amines. Supercritical CO_2 reacts with the NH_2-group to form carbamates which are not soluble in supercritical CO_2. Shielding the aminogroup *via* an acyl derivative is the recommended technique.[30]

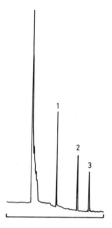

Figure 16 *CSFC–Tertiary Amines. Column:* 25 m × 0.1 mm *i.d., OV-73; temperature:*
150 °C; *fluid:* CO_2, *linear pressure programme,* 15 MPa (15 min) *to* 25 MPa
at 0.35 MPa min⁻¹. *Peaks: 1: dimethylhexadecylamine, 2: methyldihexa-*
decylamine, 3: trihexadecylamine.
(Reproduced by permission from *J. High Resolut. Chromatogr., Chroma-*
togr. Commun., to be published)[30]

4 Conclusion

High temperature capillary gas chromatography and capillary supercritical
fluid chromatography should not be considered to be competitive to each
other. Each technique has its own possibilities and is therefore complementary to the other.

HT-CGC offers the highest efficiency in the shortest analysis time. This
is not only valid *vs.* CSFC but also *vs.* HPLC. In combination with cold
on-column injection, sample discrimination as a function of molecular
weight is completely avoided. A limitation of HT-CGC is that only a few
stationary phases withstand temperatures up to 400 °C.

The main advantage of CSFC is that the compounds elute at much lower
temperatures, which is a must for certain classes of compounds, *i.e.* phospholipids. In this case, CSFC is also to be preferred over HPLC because of the
applicability of GC detectors. The low temperatures applied in CSFC will
allow the use of highly selective stationary phases for the analysis of HMW
compounds. To fully exploit the possibilities of CSFC, new developments are
necessary, especially in sample introduction, in capillary columns with i.d.'s
in the order of 10 to 20 μm and in enhancing the selectivity and solvation
power of the supercritical medium.

References

1. F. Munari, C. Saravalle, A. Della Foglia, S. Trestianu, M. Galli, J. M. Colin, and J. L. Jovelin *in* 'Proceedings of Sixth International Symposium on Capillary Chromatography', *ed.* P. Sandra, Huethig, Heidelberg, 1985, p. 417.
2. L. A. Luke and J. E. Ray, *J. High Resolut. Chromatogr., Chromatogr. Commun.*, 1985, **8**, 193.
3. S. Trestianu, G. Zilioli, A. Sironi, C. Saravalle, F. Munari, M. Galli, G. Gaspar, J. M. Colin, and J. L. Jovelin, *J. High Resolut. Chromatogr., Chromatogr. Commun.*, 1985, **8**, 771.
4. S. Trestianu, M. Galli, J. L. Jovelin, and G. D. Dupre *in* 'Proceedings of Eighth International Symposium on Capillary Chromatography', *ed.* P. Sandra, Huethig, Heidelberg, 1987, p. 851.
5. S. R. Lipsky, *LC & GC*, 1986, **4**, 898.
6. E. Geeraert, P. Sandra, and D. De Schepper, *J. Chromatogr.*, 1983, **279**, 287.
7. E. Geeraert and P. Sandra, *J. High Resolut. Chromatogr., Chromatogr. Commun.*, 1984, **7**, 431.
8. E. Geeraert and P. Sandra, *J. High Resolut. Chromatogr., Chromatogr. Commun.*, 1985, **8**, 415.
9. E. Geeraert *in* 'Sample Introduction in Capillary Gas Chromatography', Vol. 1 *ed.* P. Sandra, Huethig, Heidelberg, 1985, 159.
10. J. V. Hinshaw, Jr. and W. Seferovic, *J. High Resolut. Chromatogr., Chromatogr. Commun.*, 1986, **9**, 731.
11. E. Geeraert and P. Sandra, *J. Am. Oil Chem. Soc.*, 1987, **64**, 100.
12. M. Termonia, F. Munari, and P. Sandra, *J. High Resolut. Chromatogr., Chromatogr. Commun.*, 1987, **10**, 263.
13. Application Note, *Chromatogr. Int.*, 1986, (20), 16.
14. E. M. Martinelli, R. Seraglia, and G. Pifferi *in* 'Proceedings of Sixth International Symposium on Capillary Chromatography', *ed.* P. Sandra, Huethig, Heidelberg, 1985, p. 429.
15. G. di Pasquale, L. Giambelli, A. Soffientini, and R. Paiella, *J. High Resolut. Chromatogr., Chromatogr. Commun.*, 1985, **8**, 618.
16. W. Blum and L. Damasceno, *J. High Resolut. Chromatogr., Chromatogr. Commun.*, 1987, **10**, 472.
17. P. Sandra, unpublished results.
18. H. E. Schwartz, J. W. Higgins, and R. G. Brownlee, *LC & GC*, 1986, **4**, 639.
19. H. E. Schwartz, R. G. Brownlee, M. M. Boduszynski, and F. Su, *Anal. Chem.*, 1987, **59**, 1393.
20. T. L. Chester, *J. Chromatogr.*, 1984, **299**, 424.
21. C. M. White and R. K. Houck, *J. High Resolut. Chromatogr., Chromatogr. Commun.*, 1985, **8**, 293.
22. M. Proot, P. Sandra, and E. Geeraert, *J. High Resolut. Chromatogr., Chromatogr. Commun.*, 1986, **9**, 189.

23. P. Sandra, M. Proot, and E. Geeraert *in* 'Proceedings of Seventh International Symposium on Capillary Chromatography' *eds.* D. Ishii, K. Jinno, and P. Sandra, The University of Nagoya Press, Japan, 1986, p. 650.
24. W. P. Jackson and D. W. Later, *J. High Resolut. Chromatogr., Chromatogr. Commun.*, 1986, **9**, 175.
25. T. L. Chester and D. P. Innis, *J. High Resolut. Chromatogr., Chromatogr. Commun.*, 1986, **9**, 209.
26. S. B. Hawthorne and D. J. Miller, *J. Chromatogr.*, 1987, **388**, 397.
27. C. M. White and R. K. Houck, *J. High Resolut. Chromatogr., Chromatogr. Commun.*, 1986, **9**, 4.
28. R. L. Eatherton, M. A. Morrissey, W. F. Siems, and H. H. Hill, Jr., *J. High Resolut. Chromatogr., Chromatogr. Commun.*, 1986, **9**, 154.
29. P. Sandra, F. David, F. Munari, G. Mapelli, and S. Trestianu, *LC & GC*, submitted for publication.
30. F. David, P. Sandra, and G. Mapelli, *J. High Resolut. Chromatogr., Chromatogr. Commun.*, in the press.
31. P. Sandra and F. David, *J. High Resolut. Chromatogr., Chromatogr. Commun.*, submitted for publication.
32. M. Verzele, F. David, M. Van Roelenbosch, G. Diricks, and P. Sandra, *J. Chromatogr.*, 1983, **279**, 99.
33. Y. Hirata, F. Nakata, and M. Horihata in 'Proceedings of Eighth International Symposium on Capillary Chromatography', *ed.* P. Sandra, Huethig, Heidelberg, 1987, p. 964.
34. G. Mapelli, unpublished results.
35. S. M. Fields, R. C. Kong, J. C. Fjeldsted, and M. L. Lee, *J. High Resolut. Chromatogr., Chromatogr. Commun.*, 1984, **7**, 312.
36. M. L. Lee, lecture presented at the Eighth Int. Symp. on Cap. Chromatogr., May 19–21, Riva del Garda, Italy.

CHAPTER 6

Supercritical Fluid Chromatography-Mass Spectrometry

DAVID E. GAMES, ANTONY J. BERRY,
IAN C. MYLCHREEST, JOHN R. PERKINS, and
STEPHEN PLEASANCE

1 Introduction

Because the physical properties of a supercritical fluid resemble a gas in viscosity, a liquid in density, and are intermediate between these two phases in terms of diffusivity, it provides a favourable medium for the transport of solutes through a chromatographic column. Supercritical fluid chromatography (SFC) is complementary to both gas chromatography (GC) and high performance liquid chromatography (HPLC) for the analysis of organic compounds and, like HPLC, does not require vaporization of sample, thus enabling involatile compounds, which are not amenable to gas chromatographic study, to be analysed.

Capillary SFC, which uses fused silica columns to which heavily cross-linked bonded stationary phases have been applied, provides high chromatographic efficiency, enables use of density (pressure) programming,[1,2] and can be readily interfaced to gas chromatographic detectors.[3] This approach has the disadvantages that analysis time can be long if optimal chromatographic efficiency is sought, split injection techniques are often used, which can result in poor quantification, and sample size and injection volumes are limited. An alternative and complementary approach to SFC uses conventional packed HPLC columns.[2,4] Here, UV detection is normally used; however if a splitting of the eluent is effected, gas chromatographic detectors can be utilized. Advantages of this approach include speed of analysis and method development, improved detection limits over HPLC, ease of use of low wavelength UV detection, and preparative isolation of high purity samples. However, chromatographic efficiencies are not as good as with capillary

SFC and high percentages of polar modifiers have to be used for analysis of polar compounds which would cause problems with some types of GC detector.

Combined gas chromatography-mass spectrometry (GC-MS) is now a well established routinely used technique. High performance liquid chromatograhy-mass spectrometry (HPLC-MS) has, in recent years, become increasingly widely used, though there is as yet no universal system. The reasons for the development of these techniques are that the mass spectrometer can provide specific, selective, and sensitive detection of components eluting from a chromatograph. The approach also enables multicomponent peaks to be identified and components to be resolved, and stable isotopically labelled compounds can be used as internal standards for quantification. Similar benefits would accrue from SFC-MS.

Because the mobile phase gas flow rates generated from supercritical fluids are higher and the requirement to maintain supercritical fluid conditions in the interface, SFC-MS is not as easy to effect as GC-MS. However, the mobile phase gas flow rates are lower than are applicable in HPLC-MS and hence interfacing is easier. For effective development of an SFC-MS interface, there are three prime requirements.

(*a*) The interface must be capable of handling the gas flow rates generated from the mobile phase. This is obviously easier with capillary SFC than packed-column SFC. In the former case, it will be seen later that conventional mass spectrometers, configured for chemical ionization, are well capable of handling the flow rates. For packed-column studies, either removal of mobile phase is required or it could be used to assist in the mass spectral ionization process.

(*b*) The solute must be transported into the mass spectrometer ion source and ionization effected without thermal decomposition. Ionization in mass spectrometry can be divided into two types:-

  (*i*) Vaporization of the sample followed by ionization. The ionization techniques which are most commonly used are electron impact (EI) and chemical ionization (CI). In the former case, a high energy beam of electrons is used and in the latter case, a reagent gas (*e.g.*, methane) is ionized by use of an electron beam or a discharge, and the ions generated are used to ionize the solute. This approach is restricted to solutes which can be vaporized without thermal decomposition and, at the present state of SFC, EI and/or CI is suitable for ionization of most solutes which have been demonstrated to be amenable to SFC study.

  (*ii*) Ionization from the condensed phase. In order to effect ionization of highly polar or ionic, involatile molecules alternative ionization methods have to be used. Fast atom bombardment (FAB), where ionization is effected by bombarding a solution of the sample in a matrix (commonly glycerol) with xenon or argon neutrals or charged particles, is the most popular technique.[5] Thermospray ionization has also been shown to be very effective for the ionization of a wide range of mass spectrometrically difficult molecules and has the advantages of ease of use for HPLC-MS.[6] Other ionization methods which have proved to be useful in this context are second-ary ion mass spectrometry,[7] field desorption,[8] [252]californium plasma desorption,[9] electrospray ionization,[10] laser desorption,[12] and electrohydrodynamic ionization.[13] As far as the authors of this chapter are aware, as yet no literature reports have appeared describing SFC studies of molecules requiring these types of ioniza-tion. However, early reports on SFC indicated the ability of the technique to migrate molecules such as peptides,[14] and as SFC develops SFC-MS interfaces will probably need to be capable of ionizing polar and/or ionic involatile molecules. Consideration of HPLC-MS interfaces indicate that use of continuous flow FAB[15] or FAB with a moving belt interface,[16] electrospray ionization,[10,17] liquid ion evaporation,[11,17] and thermospray ionization[6] could be adapted for SFC-MS studies of mass spectrometrically difficult compounds.

(*c*) Maintenance of chromatographic integrity; this can be achieved by using minimal dead volume couplings.

Before proceeding to consider different approaches to SFC-MS, it is appropriate to mention some further mass spectrometric considerations, in particular mass spectrometer instrument design in terms of mass analysis. Currently, most instruments use either quadrupoles or a combination of magnetic and electric fields to effect mass analysis. The former type of instruments have lower source potentials making interfacing with chromato-graphs easier. Recent developments have considerably extended the mass range of such instruments; however, magnetic instruments have greater mass ranges, better high mass sensitivity, and enable high resolution accurate mass data to be obtained. Three other types of mass analysis are currently exciting considerable interest. These are time-of-flight mass spectrometers, which have advantages in terms of mass range and detection limits, though are currently limited in terms of resolving power. Fourier transform (FT) mass spectrometry is another rapidly developing technique which offers high mass capability, advantages in conducting collision-induced dissociation, and good detection limits.[18] However, for optimal use, very low pressures have to be maintained and hence chromatographic interfacing is difficult. Finally, ion trap mass spectral instruments offer a cheap alternative to conventional mass spectrometers and are being extensively used for GC-MS.[19] Currently, mass range and resolution are limited but there are rapid developments in the area. As with FT mass spectrometry maintenance of low source press-ures is important. As will be discussed later, SFC-MS has been effected with both FT and ion trap mass spectometers..

Several reviews of SFC-MS[20-22] have appeared in the literature and the objectives of this chapter will be to discuss the various types of SFC-MS interfaces which have been reported and to discuss areas where the technique has been applied or has potential application.

2 Capillary Supercritical Fluid Chromatography/ Mass Spectrometry

Typical fluid flow rates for capillary SFC, if 50 μm i.d. columns are used, are in the range 0.5–5 μl min^{-1} (as liquid). This flow can be adequately handled by mass spectrometers configured for CI. The first system described for capillary SFC-MS[23,24] consisted of a heated direct fluid injection probe, which carried the fused silica capillary column to within 3 cm of the probe tip, where it was connected by a zero dead volume connection to 100 μm i.d. platinum-iridium tubing in the probe tip. Two types of flow restriction were examined; one utilizing a 0.5–2.0 μm laser drilled orifice, the other using a 0.2–0.5 mm length of capillary restriction. In the latter case, restriction was effected by crimping the tubing to obtain desired flow rates. The two probes gave similar performance, but the first approach was more prone to plugging. Operation of the system in the CI mode with methane, i-butane, or ammonia as reagent gases showed that sub-nanogram, for full scan spectra, and low picogram, for selected ion monitoring (SIM), detection limits were possible for a range of polynuclear aromatic hydrocarbons. This approach, when a rapid pressure drop and proper temperatures are maintained, utilizes the shock fronts and rapid collisional processes in the expanding jet to disrupt cluster species and thus prevents droplet formation.[25] Use of short columns with rapid programming rates enabled rapid SFC-MS separations to be effected with this interface.[26] The narrow peak widths thus produced gave detection limits of 0.1 pg in SIM for a series of biphenyls. It should be noted, however, that such an approach results in considerable loss of chromatographic efficiency.

Initial attempts at obtaining EI spectra with this system resulted in spectra containing both CI and EI components and reduced sensitivity.[24] These studies were conducted with 100 μm i.d. columns. 50 μm i.d. columns have lower flow rates and modification of an EI source has been described for EI SFC-MS.[27] Formation of cluster ions was prevented by incorporation of a heated expansion region which was maintained 50–150 °C higher than the mobile phase temperature. The system provided EI type spectra without noticeable CI contributions, although sensitivity was not as good as in the CI mode. Scanning of the instrument was normally from m/z 100 because of the formation of $[CO_2]^+$ and $[(CO_2)_2]^+$ at m/z 44 and 88, respectively. Detailed studies of the effects of source pressure and temperature on the formation of CO_2 cluster ions in the EI mode have been reported.[28] In this case, a 10 μm i.d. fused silica restrictor was used and the EI source was not modified. Ionization appears to be effected both by charge exchange with CO_2 and by electron impact, the contributions from the two mechanisms being hard to differentiate. Further improvements have been effected by use of a Guthrie[29] type restrictor, more open EI source design and heating of the restrictor.[30] This modified system gave improved EI sensitivity, however it was still not as good as in the GC-MS mode.

Further improvements in capillary SFC-MS have been effected by appropriate choice of restrictors and a detailed study of a variety of restrictor types has been reported.[31] The best flow conditions for capillary SFC restrictors are obtained when relatively short restrictors are used and heating of the fluid prior to the restrictor or during decompression is necessary. In the latter case, tapered silica restrictors or frit restrictors are preferable and are also best for the transport of involatile compounds.

The capillary SFC-MS systems so far discussed have used split injection techniques. This approach can cause problems for quantification and, since only relatively small sample sizes and injection volumes can be used to avoid overloading the column, problems with solubility and dynamic range are encountered. An SFC-MS interface has recently been developed which allows much higher flow rates to be handled and which is also suitable for microbore SFC-MS.[32] The system incorporates a mechanically pumped expansion region behind the chemical ionization repeller electrode and directly heated drawn fused silica restrictors were used. Further significant advances have been the development of a capillary SFC-MS system for magnetic instruments[33] and for a bench-top mass spectrometer.[34]

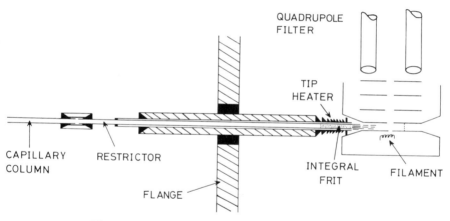

Figure 1 *Diagram of a capillary SFC-MS interface*

We have developed capillary SFC-MS interfaces for our Finnigan 4500 mass spectrometer.[21,22,35] Initially we used a 5 µm i.d. fused silica restrictor attached to the capillary SFC column with a zero dead volume connector and used the capillary GC-MS transfer line of the mass spectrometer. Problems were encountered with the handling of low volatility compounds. These have been overcome by using a modified direct introduction probe system (Figure 1). A 50 µm i.d. fused fritted silica integral restrictor was connected to the end of the capillary column *via* a zero dead volume union and a restrictor was inserted into the probe and was sealed into position such that the restrictor and probe tips were aligned. The probe was inserted into the manifold *via* a Swage vacuum lock until the tip was sited 10 mm outside the ion volume of the source and the restrictor and probe tips were directly

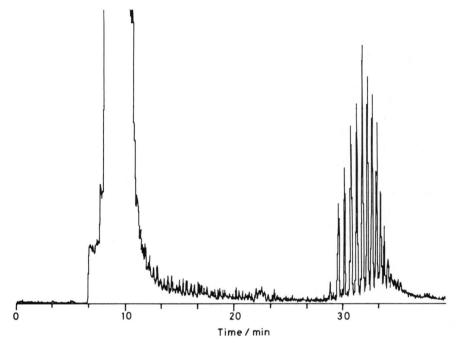

Figure 2 *Capillary EI SFC-MS computer reconstructed total ion current trace of a high b.p. alkane fraction. Column:* $15 \times 50\,\mu m$ *methyl siloxane; mobile phase: carbon dioxide; column temperature:* $100\,°C$; *density programmed from* $0.2\,g\,cm^{-3}$ *to* $0.8\,g\,cm^{-3}$ *at* $0.02\,g\,cm^{-3}\,min^{-1}$; *interface temperature:* $480\,°C$

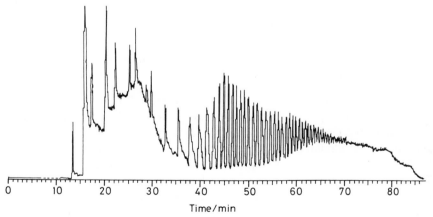

Figure 3 *Capillary methane CI SFC-MS computer reconstructed total ion current trace from polydimethyl siloxane. Column:* $10\,m \times 50\,\mu m$ *methyl siloxane; mobile phase: carbon dioxide; column temperature:* $80\,°C$; *initially held at* $0.2\,g\,cm^{-1}$ *for 10 min., then asymptotically programmed so that at* $t_{\frac{1}{2}}$ *(20 min.) the density was* $0.86\,g\,cm^{-1}$; *interface temperature:* $350\,°C$

heated to prevent solute precipitation during eluent decompression. The system enables EI type spectra and conventional CI spectra to be obtained. Best sensitivity being obtained in the CI mode. Figure 2 shows the computer reconstructed total ion current (TIC) trace obtained under EI conditions from a high boiling point mixture of alkanes. Use of methane CI provides spectra with $[M-1]^+$ ions as the base peaks enabling molecular weights to be readily assigned, and the EI spectra provide an indication of the level of branching in the molecules. Examination of non-polar involatile compounds such as polymers by capillary SFC-MS requires high mass range instruments. Recently, quadrupole instruments have been developed with mass ranges up to 5000 u. Figure 3 shows the TIC trace which we obtained from polydimethylsiloxane using methane as a CI reagent gas, using a Finnigan MAT TSQ 70 triple quadrupole mass spectrometer. Later eluting peaks showed ions in excess of m/z 1900.

We have found capillary SFC-MS particularly useful for the study of non-polar involatiles, *e.g.*, high b.p. fractions and residues from petroleum and coal, and have used packed-column SFC-MS for more polar molecules. However, capillary SFC-MS has been effectively applied in studies of tricothecene mycotoxins,[36] acid and carbamate pesticides,[37] environmental samples,[20,21] and sugars[20] and shows considerable potential for the analysis of complex mixtures of polar involatile and thermally labile compounds.

Capillary SFC-MS also offers possibilities for the provision of selective cluster ions which can be used as diagnostic ions in analytical studies. A recent study has shown that non-retained solutes in contact with non-retained reagents, *e.g.*, pyridine, toluene, di-i-propyl oxide and halogenated solvents can enable selective ionizations to be effected.[38]

Fourier transform mass spectrometry (FTMS) offers potential advantages of enhanced mass range, resolving power, and scan speeds over currently used mass analysers. However, for optimal performance, very low pressures are required to be maintained. Recently, capillary SFC-MS has been effected using a FT mass spectrometer with a differentially pumped tandem cell.[39] Interfacing was effected using a tapered fused silica capillary restrictor which was connected to the end of a 50 μm i.d. capillary column. The transfer line for GC-MS was used to house the interface. In these preliminary experiments, the transfer line was maintained at 100 °C and incorporation of selective heating for the terminal restrictor would improve the system's performance. It was shown that the system could tolerate the volumes of gas introduced from a capillary SFC system. Ability to obtain high resolution accurate mass measurements under EI and self-CI conditions was demonstrated and, unlike capillary SFC-MS with quadrupole instruments, sensitivity was comparable in the two modes. The capability of the system to perform conventional CI with i-pentane as reagent gas was also demonstrated. These experiments illustrate that, with improvements in interface design, capillary SFC-FTMS is feasible. Similar considerations in interfacing pertain to the use of an ion trap for capillary SFC-MS. Recently, we have effected interfacing to such a mass analyser.[40] The capillary SFC column

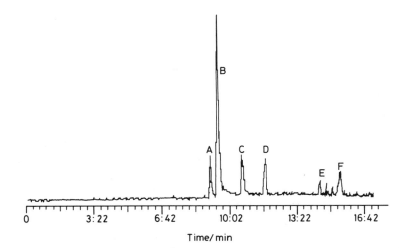

Figure 4 *Capillary SFC-MS with an ion trap mass spectrometer. Computer reconstructed total ion current trace of a mixture of benzene (B), butylbenzene (C),
naphthalene (D), anthracene (E), and pyrene (F) in dichloromethane (A).
Column: 15 m × 50 μm methyl siloxane; mobile phase: carbon dioxide;
column temperature: 75 °C; density programmed from 0.2 g cm^{-3} to
0.5 g cm^{-3} at 0.03 g cm^{-3} min^{-1}, then 0.1 g cm^{-3} min^{-1} up to 0.9 g cm^{-3};
interface: 405 °C. Helium was added to the ion trap to give an initial pressure
in the trap of 3.2 × 10^{-5} Torr and a final pressure of 8 × 10^{-4} Torr*

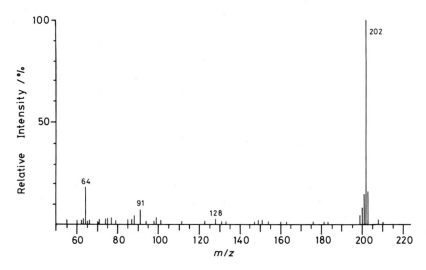

Figure 5 *Mass spectrum of pyrene obtained by capillary SFC-MS with an ion trap
detector*

was passed through the ion trap detector GC-MS transfer line and coupled *via* a zero dead volume connector to a frit restrictor. Initial experiments were conducted with carbon dioxide mobile phase only entering the trap and a series of aromatic compounds, naphthalene, butyl benzene, and phenanthrene, were injected into the SFC and produced reasonable total ion current traces and good EI spectra. Improvements in spectral quality were obtained by introduction of helium into the trap. Figure 4 shows the TIC trace obtained from a mixture of aromatic compounds and Figure 5 the spectrum obtained from the last eluting peak pyrene. Studies are in progress to effect improved pumping with the system which should enable maintenance of low pressures and also to see if more difficult molecules can be handled.

3 Packed-column Supercritical Fluid Chromatography-Mass Spectrometry

The gas volumes generated from the mobile phase, if conventional high performance liquid chromatographic columns (4.6 mm i.d.) are used, are much greater than from capillary columns and hence SFC-MS interfacing presents greater problems than in the capillary mode of operation. An alternative approach is to use narrower bore columns, which result in reduced mobile phase flows and this approach, which will be described later, has been successfully applied. An early packed-column SFC-MS system used a splitting device which fed a small portion of the column eluent into the mass spectrometer ion source *via* a capillary inlet.[41] The mass spectrometer was operated in the EI mode and the technique was successfully applied to the analysis of oil fractions. The system was not amenable to the study of lower volatility compounds and, because of the very high split ratios, detection limits were very poor. A more complex approach involved the use of a molecular beam interface.[42,43] The interface used a supersonic jet model and a skimmer to reduce Mach disk interference in forming a molecular beam. Incorporation of several stages of differential pumping provided source pressures appropriate for EI. It was shown that EI spectra could be obtained from a range of relatively volatile molecules and that solute solvent clustering phenomena were not serious problems, however, the complexity of the instrumentation and poor sensitivities obtained inhibited further development of this approach. Use of a modified direct liquid introduction HPLC-MS system has proved to be a very effective interface for packed-column SFC-MS.[44] A 3 μm diaphragm was used rather than the 5 μm diaphragm used for HPLC-MS, in order to maintain SFC conditions and the probe was maintained at 60 °C. Approximately 1 : 12 of the chromatographic eluent from 2.1 mm i.d. columns could be handled by the interface and, since methanol was used as modifier, solvent moderated CI mass spectra were obtained. Detection limits were found to be variable according to the class of molecule examined, *e.g.*, full scan spectra were obtainable from 20 ng injected cocaine[44] and rantidine-*N*-oxide.[45]

The capability of the system was demonstrated with a range of polar drugs and applied to the identification of phenylbutazone and its metabolites in a crude acid extract of equine urine. This system provides a practical approach to packed-column SFC-MS. Limitations include the inability to handle all the eluent from the SFC unless small bore columns are used, the inability of the diaphragm to withstand protracted use at pressures above 200 bar, and the provision of only solvent moderated CI spectra. The spectra obtained in general consist almost exclusively of protonated molecular ions. However, use of tandem or hybrid mass spectrometers, *e.g.*, triple quadrupole instruments with collisional induced dissociation can provide more structural information and give more specific identification.

Use of smaller bore high performance liquid chromatographic columns for packed-column SFC can enable all the eluent from the SFC to be handled by the mass spectrometer. A system in which a mechanically pumped expansion region behind the chemical ionization repeller electrode was incorporated and directly heated drawn fused silica restrictors were used, has been described for 1 mm i.d. columns.[32] Packed-column SFC-MS at flow rates of 60–100 µl min^{-1} have been directly handled using a thermospray source fitted with a discharge.[46] Interfacing was effected using a fused silica restrictor and solvent mediated CI mass spectra were obtained from a range of polar compounds. Since 2.1 mm i.d. columns were used at flow rates of 350 µl min^{-1}, a split was used to enable the system to handle the eluent from the SFC. Use of packed fused silica columns (0.5 mm i.d.) provides much lower column eluent flow rates, 40 µl min^{-1}. This approach has been used for SFC-MS with a self-spouting and vacuum nebulising assisted interface.[47] The system could not handle all the flow and approximately half the eluent was vented. Styrene oligomers, triglycerides, polyethylene glycol derivatives,[47] and a range of water soluble vitamins[48] have been shown to be amenable to analysis using the system. Spectra obtained were of the solvent moderated CI type.

In our studies of packed-column SFC-MS, we have concentrated on the development of systems which can be effectively used with conventional, *i.e.* 4.6 mm i.d., columns whereby a high proportion of the column eluent can be directly handled by the interface. Moving belt HPLC-MS interfaces have the advantage of providing both conventional EI and CI mass spectra and we have interfaced packed-column SFC to both VG Analytical and Finnigan MAT moving belt HPLC-MS systems.[49,50] In these studies, interfacing was effected using a modified Finnigan MAT thermospray spray deposition device. The spray deposition device was connected to our Hewlett-Packard 1084B chromatograph using a T-piece between the UV detector exit and the outlet back pressure regulator. This provides for 100% of the column eluent passing through the UV detector and approximately 50% passing through the spray deposition device. In order to maintain supercritical conditions, the end of the deposition device was crimped and heated so that no freezing of the mobile phase occurred at the tip. The system enables conventional EI mass spectra over the full scan range to be obtained and CI mass spectra.

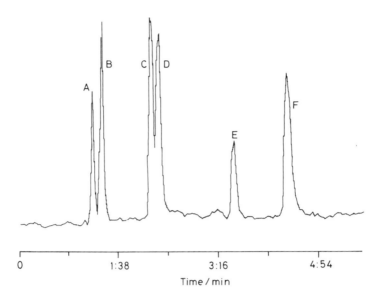

Figure 6 *Computer reconstructed total ion current trace obtained by EI packed column SFC-MS from a mixture of atrazine (A), simazine (B), dimethirimol (C), terbacil (D), menazon (E), and ethirimol (G). Column: 100 × 4.6 mm amino bonded Spherisorb (5μm); mobile phase: carbon dioxide and methanol programmed from 5% to 20% methanol over 7 min at a flow of 4 ml min⁻¹; column pressure: 275 bar; column temperature: 90 °C*

Because the peaks obtained by packed-column SFC are narrower than from HPLC and there is less background, detection limits are improved over use of such systems in the HPLC-MS mode. A wide range of compounds, including sulphonamides, steroids, coumarin rodenticides, indole alkaloids, xanthines, PTH amino acids, veterinary drugs, and milbemycin antibiotics have been shown to be amenable to study with the system. Figure 6 shows the TIC trace obtained by packed-column EI SFC/MS using a moving belt interface from a mixture of the pesticides atrazine, simazine, dimethirimol, terbacil, menazon, and ethirimol. Table 1 summarises the EI mass spectra obtained from these compounds and the TIC trace is in good agreement with the UV trace obtained simultaneously with the TIC trace.

Our second approach to packed-column SFC-MS has utilized a Finnigan MAT thermospray source in the filament-on mode. Initially the coiled stainless steel hypodermic tubing used for the vaporizer was replaced by a straight piece of this tubing and the end was crimped to maintain SFC conditions until the eluent left the vaporizer.[51] Figure 7 shows a schematic of the modified source. Recently, we have obtained better performance using a conventional coiled vaporizer with the end crimped. In this mode, if a modifier is used, solvent mediated CI mass spectra are obtained and Table 1 gives the spectra obtained in this mode from the mixture of six pesticides run by packed-column SFC-MS and with the belt interface (Figure 6).

Table 1 *Mass spectra obtained from a pesticide mixture by SFC-MS using a moving belt and filament-on thermospray interfaces.*

Peak	Compound	Mol. wt.	Belt m/z (% rel. int.)	Thermospray m/z (% rel. int.)
A	atrazine	215	217(18), 215(56), 202(30), 200(100), 173(29), 92(18), 64(41), 58(67)	218(32), 216(100)
B	simazine	201	203(32) 201(100), 188(18), 186(59), 173(42), 158(26), 71(38), 68(73)	204(33), 202(100)
C	dimethirimol	209	209(17), 166(100), 110(2), 96(5), 71(8), 55(7)	210(100)
D	terbacil	216	218(2), 216(5), 164(26), 163(25), 161(100),160(71), 117(42), 57(43)	219(19), 217(57), 195(31), 193(100)
E	menazon	281	281(9), 156(100), 125(27), 93(28)	282(100)
F	ethirimol	209	209(18), 166(100), 138(5), 96(26), 55(13)	210(100)

All the compounds showed $[M + 1]^+$ ions with little or no fragment ions. This lack of structural information can present problems and we have recently found that variation of the repeller voltage in the ion source can induce the production of spectra with fragment ions. An alternative approach to the production of more structural information would be the use of a hybrid or tandem instrument with collisional induced dissociation.

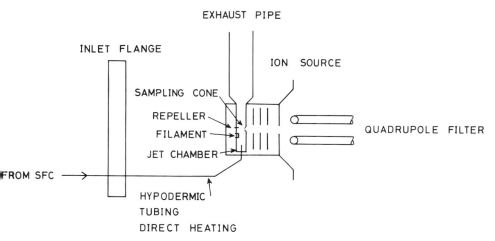

Figure 7 *Diagram of filament-on thermospray SFC-MS system*

Alternatively, if only carbon dioxide can be used as the mobile phase, charge EI type spectra are produced.[51] Detection limits in this mode are low nanogram → high picogram for full scan spectra with low picogram detection, if selected ion monitoring is used. However, because of the presence of cluster ions, scanning is normally commenced from above 120 u. We have found this interface to be useful for the study of a wide range of compound types, including veterinary drugs,[22] sulphonamides, organophosphorous and carbamate pesticides, indole alkaloids, PTH amino acids, and steroids.

4 Conclusions

A range of effective interfaces has been developed for both packed and capillary SFC-MS. Because, as yet, the range of compounds shown to be amenable to SFC is limited, currently developed interfaces can provide mass spectral information from most classes of compound. Capillary SFC-MS provides high efficiency separation and high sensitivity if used in the CI mode. EI type spectra can be obtained with this mode of operation but with poorer detection limits. Whilst this is the approach of choice for the analysis of very complex mixtures, packed-column SFC-MS has advantages in terms of speed of analysis and speed of method development when simpler mixtures are examined. It also has the advantage that, if moving belt interfaces are used, true EI spectra are obtained. Use of a MAGIC HPLC-MS interface[52] should also enable similar information to be provided. As and when mass spectrometrically more difficult molecules, *e.g.*, underivatised peptides and polysaccharides, are shown to be amenable to SFC study, HPLC-MS interfaces such as the continuous flow FAB system and thermospray system should prove to be amenable to adaption for SFC-MS to handle these more difficult molecules.

Acknowledgements

We thank the SERC, AFRC, and Ministry of Agriculture, Fisheries and Food (MAFF) for financial assistance in the purchase of SFC and mass spectral equipment. Financial support for A. J. Berry (SERC and BP Research), I. C. Mylchreest (Glaxo and Finnigan MAT), J. R. Perkins (MAFF), and S. Pleasance (SERC and Dalgety UK) is gratefully acknowledged. We are also indebted to Finnigan MAT, Hewlett-Packard and VG Analytical for their support of our studies in this area.

References

1. P. A. Peaden and M. L. Lee, *J. Liq. Chromatogr.*, 1982, **5** (Suppl. 2), 179.
2. C. M. White and R. K. Houck, *J. High Resolut. Chromatogr., Chromatogr. Commun.*, 1986, **9**, 4.
3. M. Novotny, *J. High Resolut. Chromatogr., Chromatogr. Commun.*, 1986, **9**, 137.
4. D. R. Gere, *Science*, 1983, **222**, 253.
5. M. Barber, R. S. Bardoli, G. J. Elliott, R. D. Sedgwick, and A. N. Tyler, *Anal. Chem.*, 1982, **54**, 645A.
6. C. R. Blakley, J. J. Carmody, and M. L. Vestal, *Anal. Chem.*, 1980, **52**, 1636.
7. A. Eicke, W. Sichtermann, and A. Benninghoven, *Org. Mass Spectrom.*, 1980, **15**, 289.
8. G. W. Wood, *Tetrahedron*, 1982, **9**, 1125.
9. R. D. Macfarlane, *Anal. Chem.*, 1983, **55**, 1247A.
10. C. M. Whitehouse, R. N. Dreyer, M. Yamashita, and J. B. Fenn, *Anal. Chem.*, 1985, **57**, 675.
11. B. A. Thomson, J. V. Iribarne, and P. J. Dziedzic, *Anal. Chem.*, 1982, **54**, 2219.
12. R. J. Cotter, *Anal. Chem.*, 1984, **56**, 485A.
13. S.-T. F. Lai and C. A. Evans, *Biomed. Mass Spectrom.*, 1979, **6**, 10.
14. J. C. Gidding, M. N. Myers, L. McLaren, and R. A. Keller, *Science*, 1968, **162**, 67.
15. Y. Ito, T. Takeuchi, D. Ishii, and M. Goto, *J. Chromatogr.*, 1985, **346**, 161.
16. J. G. Stroh, J. Carter Cook, R. M. Milberg, L. Brayton, T. Kihara, Z. Huang, K. L. Rinehart, Jr., and I. A. S. Lewis, *Anal. Chem.*, 1985, **57**, 985.
17. T. R. Covey, E. D. Lee, A. P. Bruins, and J. D. Henion, *Anal. Chem.*, 1986, **58**, 1451A.
18. A. G. Marshall, *Acc. Chem. Res.*, 1985, **18**, 316.
19. G. C. Stafford, Jr., P. E. Kelley, and D. C. Bradford, *Am. Lab.*, 1983, **15** (June), 51.
20. B. W. Wright, H. T. Kalinoski, H. R. Udseth, and R. D. Smith, *J. High Resolut. Chromatogr., Chromatogr. Commun.*, 1986, **9**, 145.

21. D. E. Games, A. J. Berry, I. C. Mylchreest, J. R. Perkins, and S. Pleasance, *Lab. Pract.*, 1987, **36(2)**, 45.
22. D. E. Games, A. J. Berry, I. C. Mylchreest, J. R. Perkins, and S. Pleasance, *European Chromatography News*, 1987, **1(1)**, 10.
23. R. D. Smith, W. D. Felix, J. C. Fjeldsted, and M. L. Lee, *Anal. Chem.*, 1982, **54**, 1883.
24. R. D. Smith, J. C. Fjeldsted, and M. L. Lee, *J. Chromatogr.*, 1982, **247**, 231.
25. R. D. Smith and H. R. Udseth, *Anal. Chem.*, 1983, **55**, 2266.
26. R. D. Smith, H. T. Kalinoski, H. R. Udseth, and B. W. Wright, *Anal. Chem.*, 1984, **56**, 2476.
27. R. D. Smith, H. R. Udseth, and H. T. Kalinoski, *Anal. Chem.*, 1984, **56**, 2971.
28. G. Holzer, S. Deluca, and K. J. Voorhees, *J. High Resolut. Chromatogr., Chromatogr. Commun.*, 1985, **8**, 528.
29. E. J. Guthrie and H. E. Schwartz, *J. Chromatogr. Sci.*, 1986, **24**, 236.
30. S. D. Zaugg, S. J. Deluca, G. U. Holzer, and K. J. Voorhees, *J. High Resolut. Chromatogr., Chromatogr. Commun.*, 1987, **10**, 100.
31. R. D. Smith, J. L. Fulton, R. C. Petersen, A. J. Kopriva, and B. W. Wright, *Anal. Chem.*, 1986, **58**, 2057.
32. R. D. Smith and H. R. Udseth, *Anal. Chem.*, 1987, **59**, 13.
33. H. T. Kalinoski, H. R. Udseth, E. K. Chess, and R. D. Smith, *J. Chromatogr.*, 1987, **394**, 3.
34. E. D. Lee and J. D. Henion, *J. High Resolut. Chromatogr., Chromatogr. Commun.*, 1986, **9**, 172.
35. A. J. Berry, D. E. Games, I. C. Mylchreest, J. R. Perkins, and S. Pleasance, *J. High Resolut. Chromatogr., Chromatogr. Commun.*, in press.
36. R. D. Smith, H. R. Udseth, and B. W. Wright, *J. Chromatogr. Sci.*, 1985, **23**, 192.
37. H. T. Kalinoski, B. W. Wright, and R. D. Smith, *Biomed. and Environ. Mass Spectrom.*, 1986, **13**, 33.
38. P. J. Arpino and J. Cousin, *Rapid Commun. Mass Spectrom.*, 1987, **1**, 29.
39. E. D. Lee, J. D. Henion, R. B. Cody, and J. A. Kinsinger, *Anal. Chem.*, 1987, **59**, 1309.
40. J. F. J. Todd, I. C. Mylchreest, A. J. Berry, D. E. Games, and R. D. Smith, unpublished work.
41. T. H. Gouw, R. E. Jentoft, and E. J. Gallegos, *in* 'High Pressure Science and Technology' *ed.* K. D. Timmerhaus and M. S. Barber, Plenum Press, New York, 1977, Vol. 1, p. 583.
42. L. G. Randall and A. L. Wahrhaftig, *Anal. Chem.*, 1978, **50**, 1703.
43. L. G. Randall and A. L. Wahrhaftig, *Rev. Sci. Instrum.*, 1981, **52**, 1283.
44. J. B. Crowther and J. D. Henion, *Anal. Chem.*, 1985, **57**, 2711.
45. J. B. Crowther, T. R. Covey, D. Silvestre, and J. D. Henion, *LCMagazine*, 1985, **3**, 240.

46. P. Dätwyler, H. T. Walther, and P. Hirter, *4th Symposium on Liquid Chromatography Mass Spectrometry LC/MS and MS/MS,* Montreux, October 1986.
47. K. Matsumoto, S. Tsuge, and Y. Hirata, *Anal. Sci.,* 1986, **2**, 3.
48. K. Matsumoto, S. Tsuge, and Y. Hirata, *Chromatographia,* 1986, **21**, 617.
49. A. J. Berry, D. E. Games, and J. R. Perkins, *J. Chromatogr.,* 1986, **363**, 147.
50. A. J. Berry, D. E. Games, and J. R. Perkins, *Anal. Proc.,* 1986, **23**, 451.
51. A. J. Berry, D. E. Games, I. C. Mylchreest, J. R. Perkins, and S. Pleasance, *Biomed. and Environ. Mass Spectrom.,* in press.
52. A. R. C. Willoughby and R. F. Browner, *Anal. Chem.,* 1984, **56**, 2626.

CHAPTER 7

SFC-MS in the Pharmaceutical Industry

STEPHEN J. LANE

1 Introduction

Modern drug substances are commonly non-volatile and thermally or chemically labile, therefore chromatographic analysis by HPLC is necessary, rather than by GLC. In SFC the conditions of analysis are mild and no volatilisation is required, so it may also be able to handle similar drug substances, and the ability to couple supercritical fluid chromatography to mass spectrometry (SFC-MS) would provide an important alternative to HPLC-MS or GC-MS.

In recent years, the importance of routine on-line HPLC-MS for the analysis of non-volatile and labile compounds in complex matrices has become paramount to the pharmaceutical industry. However, in comparison to GLC-MS, the practical problems associated with the interfacing of HPLC to MS proved considerable, due to the basic incompatibility of the two techniques, although significant progress has been made in the past decade.[1] In our laboratory, areas of HPLC-MS application have involved the identification and characterisation of unknown products and impurities in a variety of complex matrices, including crude fermentation broths, reaction mixtures, and drug substances. To perform these analyses both thermospray and moving belt HPLC-MS systems have been employed.[1]

The major advantage of the moving belt HPLC-MS interface is its ability to provide the analyst with more than one ionization mode, allowing a choice between electron impact ionization (EI) and chemical ionization (CI). This is a desirable feature when analysing unknowns by mass spectrometry as the molecular weight information provided by positive or negative CI combined with the complementary structural information provided by EI will give a sound basis for complete or partial structural elucidation of an unknown.

In the past, a significant time has been spent in our laboratory perfecting the methodology associated with HPLC-MS *via* the moving belt interface. A typical method using the moving belt interface would involve 2 mm i.d.

HPLC columns with flow rates of up to 500 µl min^{-1} for a 50 : 50 CH$_3$CN/ H$_2$O mobile phase composition. Deposition of the eluent onto the belt would be *via* a 'home-made' heated spray depositor based on that described by Karger *et al.*[2] This approach works well, providing the required separation can be achieved without deviating too far from the above guidelines. The use of this methodology normally involved scaling down an original analytical method that used 4.6 mm i.d. HPLC columns, flow rates of 1–2 cm^3 min^{-1}, and possibly buffers, without sacrificing chromatographic fidelity. Generally this has been possible, but often results in a compromise.

The advent of SFC for the separation of drug substances provided an alternative approach which might avoid some of these problems. The technique of SFC-MS using packed columns is uniquely suited to the moving belt HPLC-MS interface by virtue of the mobile phase compositions employed, *i.e.*, CO$_2$ with polar modifiers. A large proportion of the eluent from the SFC chromatograph, utilizing conventional 4.6 mm i.d. columns and flow rates of 1–3 cm^3 min^{-1}, can be handled by the interface *via* a spray deposition device.[3]

The chromatographically favourable physical properties of supercritical fluids, namely solubility, density, diffusivity, and viscosity, have been documented,[4-6] but the range of the compounds amenable to SFC analysis remains undefined.

This chapter describes our 'in-house' construction of an SFC chromatograph and MS introduction system by modification of commercially available HPLC and MS hardware, and examples of the applications of SFC and SFC-MS in the pharmaceutical industry.

2 Design and Operation of the SFC Chromatograph

A schematic diagram of the assembled SFC chromatograph is presented in Figure 1. The instrument is based on a Gilson gradient HPLC system incorporating two 303 pumps with 10 sc heads and an Apple IIe system controller. One pump delivers liquid CO$_2$ from a cylinder with a dip tube and the second pump delivers methanol or 2-methoxyethanol as the modifier. To prevent cavitation of the carbon dioxide during the refill cycle and to ensure efficient and reproducible flow rates of CO$_2$, the head of the CO$_2$ pump is cooled to below 0 °C (Pump B, Figure 1). This is accomplished by circulating a coolant through a clamped stainless steel jacket in thermal contact with the pump head. The column oven is a modified Varian Aerograph 2700 gas chromatograph, and a Waters 490 programmable multi-wavelength detector is employed to detect the analytes on-line as they elute from the column. Sample introduction is accomplished with a Rheodyne 7125 injection valve, fitted with a 20 µl loop. The pressure in the system can be monitored by two 0–10,000 psi in-line pressure gauges and the system back-pressure is maintained and adjusted by a 200–10,000 psi back-pressure

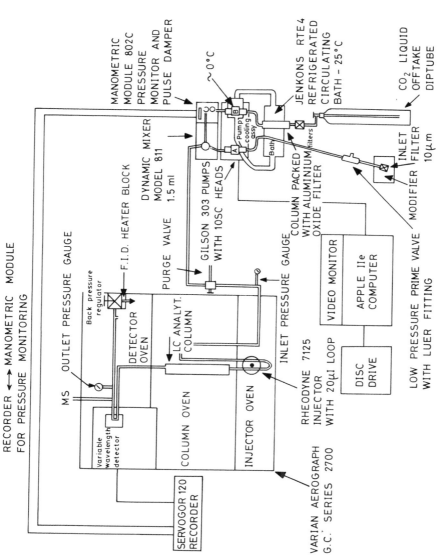

Figure 1 *Schematic diagram of the supercritical fluid chromatograph*

regulator situated downstream from the detector. A coil of $\frac{1}{16}$ inch stainless steel tubing is contained in the oven prior to the injector to facilitate the mobile phase reaching an equilibrated supercritical state before entering the injector and column. The injector was kept above the critical temperature of the mobile phase by mounting it such that the body was housed inside the oven in close proximity to the GLC injector block, which was maintained at 175 °C. The back-pressure regulator was mounted on a FID heater block and kept at a temperature high enough to prevent freezing of the mobile phase upon expansion. The separations were carried out on a 150 × 4.6 mm i.d. column, packed with 5 µm amino bonded Spherisorb or 5 µm Microsorb Si.

No permanent alterations have been required to the pumps or oven and the system can be used interchangeably for conventional HPLC or SFC. The system is capable of operating in four SFC elution modes:
(a) Isocratic-isobaric
(b) Solvent gradient
(c) Flow gradient
(d) Combined solvent-flow gradient

Coupling the SFC to the Mass Spectrometer

The SFC chromatograph was coupled to a Finnigan MAT 4600 quadrupole mass spectrometer which was equipped with a moving belt HPLC-MS interface. Interfacing of the SFC chromatograph was effected using a modified Finnigan MAT thermospray deposition device, which was connected in-line *via* a T-piece between the UV detector exit and the back-pressure regulator, thus effecting a split of eluent. To use the full density range for SFC, the end of the vaporizer tube in the spray deposition device was crimped, and the vaporizer was heated to prevent freezing of the mobile phase at the tip during expansion of the gas. The system could therefore provide both a UV and MS response sequentially on the SFC eluent.

(1)

	R^1	R^2	X—Y
Avermectin A_{1a}	ME	Bu^s	CH=CH
Avermectin A_{1b}	ME	Pr^i	CH=CH
Avermectin A_{2a}	ME	Bu^s	CH^2—CH(OH))
Avermectin A_{2b}	ME	Pr^i	CH^2—CH(OH)
Avermectin B_{1a}	H	Bu^s	CH=CH
Avermectin B_{1b}	H	Pr^i	CH=CH
Avermectin B_{2a}	H	Bu^s	CH^2—CH(OH)
Avermectin B_{2b}	H	Pr^i	CH^2—CH(OH)

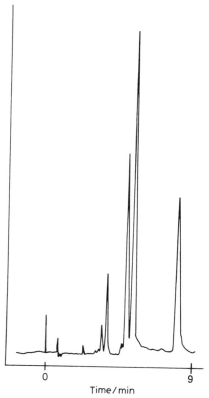

Time/min

Figure 2 *UV chromatogram for SFC separation of 22,23-dihydroavermectin B_{1a} aglycone, monosaccharide and disaccharide. Column: Rainin Microsorb Si, 5 μm, 150 × 4·6 mm; mobile phase: 7% $CH_3O(CH_2)_2OH/CO_2$, flow: 3 cm^{3}min^{-1}; pressure: 3100 psi; detector wavelength: 238 nm; 0.1 AUFS, oven temperature 65 °C*

3 Applications to Drug Substances

The SFC system was originally evaluated 'off-line' from the mass spectrometer and an early separation achieved with the system resolved the aglycone, monosaccharide and disaccharide of 22,23-dihydroavermectin B_{1a}. The UV chromatogram is shown in Figure 2. The avermectins (1) are a group of potent, broad-spectrum antiparasitic agents[7] and structurally are α-L-oleandrosyl-α-L-oleandroside derivatives of pentacyclic 16-membered lactones related to the milbemycins.[8]

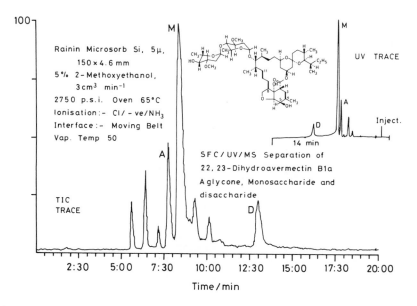

Figure 3 *Total ion current trace for the SFC negative ion chemical ionization* NH_3 *separation of 22,23-dihydroavermectin* B_{1a} *aglycone, monosaccharide and disaccharide. The in-line UV trace is inset for comparison. Peak A = aglycone, M = monosaccharide, D = disaccharide*

Subsequently, the separation has been achieved 'on-line' to the mass spectrometer, utilizing negative ion CI with ammonia as the reagent gas. Figure 3 shows the computer-reconstructed total ion current trace (TIC trace) compared to the in-line UV trace. Spectra for the three peaks, where peak A = aglycone, peak M = monosaccharide and peak D = disaccharide, gave molecular weight information as well as diagnostic fragments regarding the structure and number of sugar units present. Figure 4 shows spectra for the three peaks. SFC-EIMS gave diagnostic ions for the 16-membered lactone, but weak molecular weight information.

SFC is particularly amenable to macrolide antibiotics and, when coupled

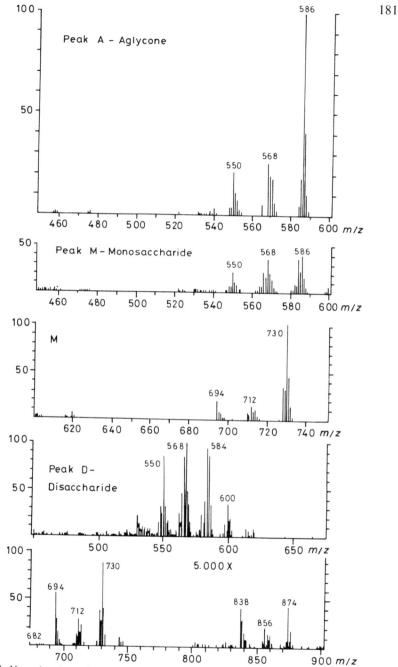

Figure 4 *Negative ion chemical ionization spectra for the three peaks of 22,23-dihydroavermectin* B_{1a} *aglycone monosaccharide and disaccharide, using ammonia as the reagent gas. Peak A = aglycone—$M^{-\cdot}$ at 586. Peak M = monosaccharide—$M^{-\cdot}$ at m/z 730 and $[M - C_7H_{12}O_3]^{-\cdot}$ at m/z 686. Peak D = disaccharide—$M^{-\cdot}$ at m/z 874 and $[M - C_7H_{12}O_3]^{-\cdot}$ at m/z 730 and $[M - 2 \times C_7H_{12}O_3]^{-\cdot}$ at m/z 586*

to mass spectrometry, it is an extremely useful technique for the identification of low level impurities present in the macrolide antibiotic of interest. Figure 5 shows the in-line UV trace of an impurity profile for a typical macrolide antibiotic.

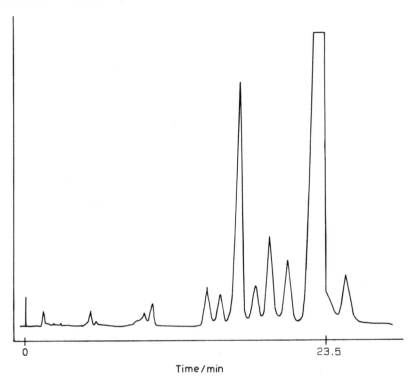

Figure 5 *UV in-line chromatogram for SFC impurity profile of a macrolide antibiotic. Column: Spherisorb* NH_2, 5 μm, 150 × 4.6 mm; *mobile phase:* 2% $CH_3OH/$ CO_2 *in* 2.5 cm³ min⁻¹; *pressure:* 4050 psi; *oven temperature:* 50 °C; *detector wavelength:* 238 nm

Chromatographic integrity is maintained through to the mass spectrometer and Figure 6 shows the TIC trace for the SFC-EIMS impurity profile. The data obtained from this run enabled structures for all the main peak related impurities to be elucidated. Figure 7 shows the EI spectrum of the main peak, and Figure 8 shows the EI spectrum of a related impurity eluting at a retention time of 19 minutes 46 seconds (see Figure 6).

Spectroscopic data obtained on a macrolide antibiotic isolated from a fermentation broth of a mutant organism suggested a novel and unexpected structure. Figure 9 shows the EI spectrum obtained *via* direct insertion probe (DIP) with a weak molecular ion being apparent at m/z 554. Isolation from the broth had utilized a reverse phase HPLC system followed by numerous harsh clean-up procedures and it was thought that the isolated component was possibly an artefact produced during isolation. To determine whether

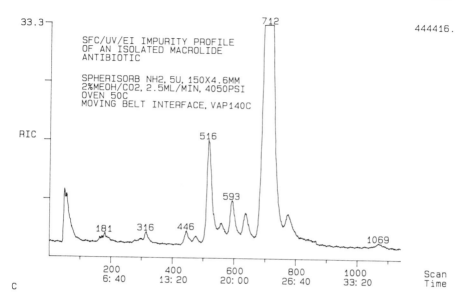

Figure 6 *Total ion current trace of the SFC-MS electron impact ionization impurity profile of a macrolide antibiotic*

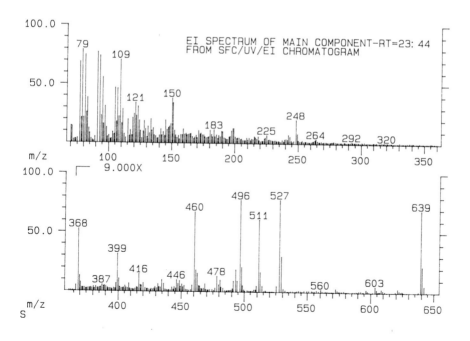

Figure 7 *Electron impact ionization spectrum of main peak (from Figure 6) showing the molecular ion at* m/z *639 and diagnostic fragments throughout the mass range*

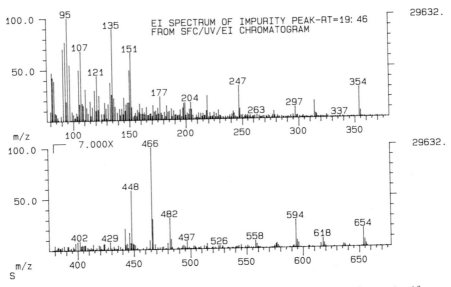

Figure 8 *Electron impact ionization spectrum of a main peak related impurity (from Figure 6) showing the molecular ion at* m/z *654 with diagnostic fragments throughout the mass range*

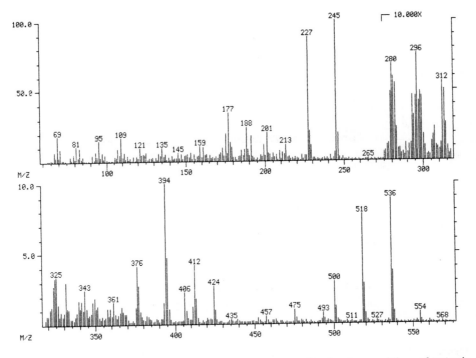

Figure 9 *Electron impact ionization mass spectrum of isolated macrolide antibiotic via direct insertion probe*

the component was present in the original broth, the fermentation was repeated and, after a minimum clean-up procedure, the methanol extract of the centrifuged cells was subjected to on-line SFC-MS in the EI and negative CI-ammonia modes of ionization. The SFC-MS results showed the component to be present in the extract. Figure 10 shows the in-line UV trace for the SFC-EIMS separation. The method involved a gradient of 3%→25% methanol/CO_2 in 25 minutes.

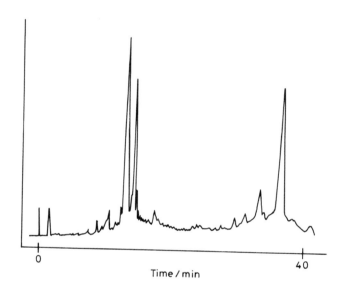

Figure 10 *In-line UV trace for gradient SFC separation of components in a crude fermentation broth extract. Column: Spherisorb NH_2, 5 μm, 150 × 4.6 mm; gradient: 3%–25% CH_3OH/CO_2 in 25 min, held 15 min, 2 ml min^{-1}; average pressure 3850 psi; oven temperature: 50 °C; detector wavelength: 238 nm*

Figure 11 shows the TIC trace under the same conditions and the selected ion (SIM) plot of characteristic ions m/z 245 and m/z 536 (Figure 12) showed that the chromatographic peak at a retention time of 15 minutes 8 seconds corresponded to the original isolated component. A full spectrum of the peak is shown in Figure 13, and it can be seen that the molecular ion at m/z 554 is not strong enough to be considered confirmatory but when the extract was run under the same conditions in the negative ion chemical ionization mode, a molecular weight of 554 was confirmed. Figure 14 shows the SIM plot monitoring m/z 554, the $M^{-\bullet}$ for the component. Figure 15 shows the full spectrum where m/z 554 is the base peak.

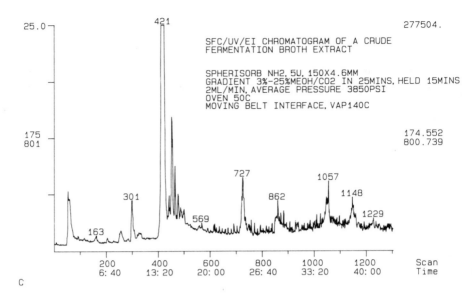

Figure 11 *Total ion current trace of the gradient SFC-MS electron impact ionization separation of components in a crude fermentation broth extract*

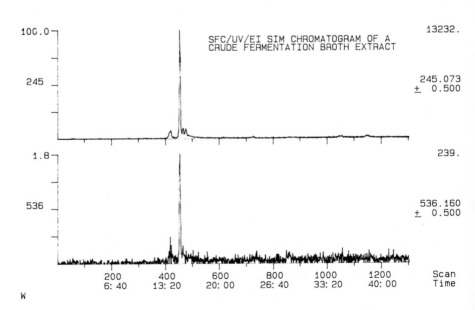

Figure 12 *Selected ion monitoring electron impact ionization plot for SFC-MS of a crude fermentation broth extract at* m/z *245 and 536 (from Figure 11)*

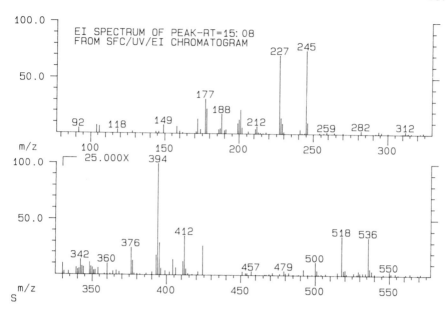

Figure 13 *Electron impact ionization mass spectrum of the peak eluting at a retention time of* 15 minutes 8 seconds (*from Figure* 12)

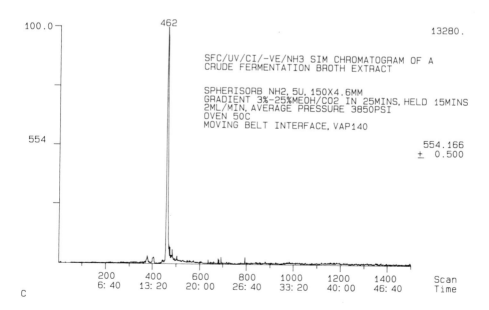

Figure 14 *Selected ion monitoring plot of the* $M^{-\cdot}$ *species at* m/z 554 (*from Figure* 12)

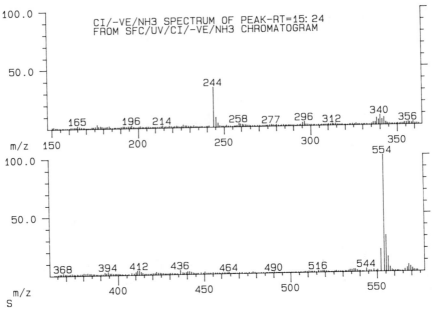

Figure 15 *Negative chemical ionization ammonia mass spectrum showing the* M⁻ᐧ *species at* m/z 554 *as the base peak (from Figure* 14)

The chromatographic separation of diastereoisomers of a cephalosporin ester (2) was easily achieved by SFC as shown in Figure 16. An impurity in the form of a structural isomer (Peak C) is well resolved from the two diastereoisomers (Peaks A and B) whilst maintaining the resolution between the diastereoisomers. When the separation is transferred to on-line SFC-MS (Figure 17), the chromatographic fidelity is maintained and characteristic spectra can be obtained. Figure 18 shows the EI spectrum for peak C.

(2)

A late-running impurity in the reversed phase HPLC of the cephalosporin sodium salt (3) persisted through subsequent reaction into the production of the ester (2) where it appeared as a broad triplet of peaks centred at a retention time of *ca.* 26 minutes. The impurity peaks were thought to be

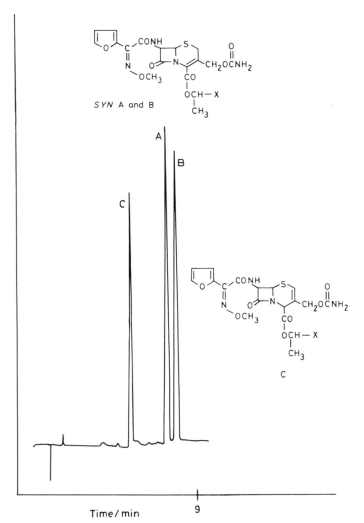

Figure 16 *UV trace for the separation of diastereoisomers (peaks A and B) and a structural isomer (peak C) of a cephalosporin ester (cefuroxime ester E47). Column: Rainin Microsorb Si, 5 μm, 150 × 4.6 mm; mobile phase: 4.5% CH₃OH/CO₂, 4.5 cm³ min⁻¹; pressure 3200 psi; detector wavelength: 276 nm; 0.64 AUFS oven temperature 55 °C*

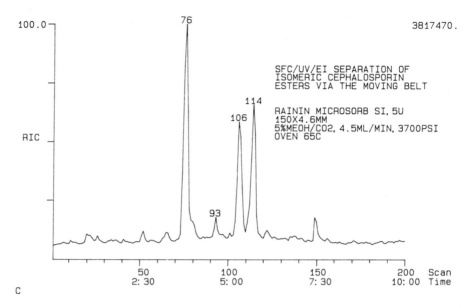

C

Figure 17 *SFC-MS separation of diastereoisomers and a structural isomer of a cephalosporin ester (from Figure 16)*

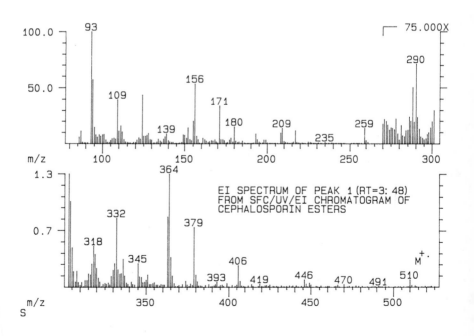

S

Figure 18 *Electron impact ionization spectrum of peak C (from Figure 16)*

diastereoisomers. The peaks amounted to *ca.* 1% in total and when sub-jected to SFC-MS a sharpened and resolved triplet eluted at a retention time of *ca.* 8 minutes. Figure 19 shows a comparison of the HPLC and SFC traces. Chromatographic integrity was maintained in the SFC-MS chroma-togram, Figure 20, and spectra were acquired over the three individual peaks using three modes of ionization, namely EI, negative CI-NH$_3$ and positive CI-NH$_3$. In each mode, spectra for the three peaks were identical and also similar to the cephalosporin ester main peak, except for the molecular ion region. Figure 21 shows the negative ion chemical ionization spectrum for the cephalosporin main peak. Figure 22 shows the spectrum for the middle impurity peak at a retention time of 8 minutes 8 seconds which was typical for all three. From these results and results obtained on the sodium salt impurity the peaks were considered to be diastereoisomers but on-line SFC-MS using EI and CI failed to give molecular weight information.

A portion of the chromatogram, Figure 20, between 6 minutes 40 seconds and 10 minutes, containing the three peaks, was collected by semi-prepara-tive SFC-UV using the standard analytical system, with conditions as for Figure 17. Ten runs were bulked leaving the analyte entrained in a small volume of methanol after evaporation of the CO$_2$. The methanol was removed by rotary evaporation and the dry analyte, after dissolution in a small volume of methanol, was subjected to fast atom bombardment (FAB)[9] in the negative ion mode using 3-nitrobenzyl alcohol as the matrix.[10] Although weak, due to the small amount of material collected, an ion at *m/z* 1045 corresponding to the [M-H]$^-$ species was consistently found, as shown in Figure 23. This evidence, together with evidence obtained on the sodium salt, allowed the structure for the impurity to be postulated.

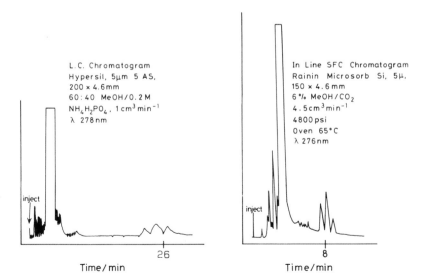

Figure 19 *Comparison of the HPLC and SFC traces for the separation of cephalo-sporin ester isomeric impurities*

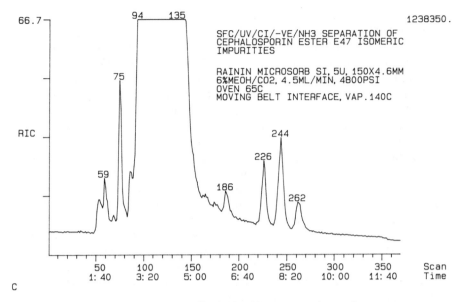

Figure 20 *Total ion current trace for SFC-MS negative chemical ionization in ammonia separation of cephalosporin ester showing three peaks of interest at a retention time of* ca. 8 minutes

The extent of ^{2}H incorporation in the Corey lactone (4) is normally determined by mass spectrometry. An attempted deuteration of (4) alpha to the lactone carbonyl gave a mixed product showing no deuterium incorporation. Figure 24 shows the EI spectrum by direct insertion probe. The molecular ion for the Corey lactone (4) is m/z 516 and the isotope pattern is unaltered but a strong ion at m/z 527 is evident. The standard Corey lactone (4) was chromatographed by SFC-EIMS; Figure 25 and Figure 26 show the EI spectra obtained. The mixed product was chromatographed under the same conditions and Figure 27 shows a comparison of the in-line UV traces for the standard and mixed product. The separation was maintained onto the mass spectrometer and Figure 28 shows the total ion current trace with the unknown product eluting at 11 minutes 24 seconds and the starting material at 7 minutes 36 seconds. The spectrum of the unknown product, Figure 29, showed the molecular weight to be 620 and this information, together with diagnostic fragments, allowed a structure to be postulated and the reaction steps rationalised.

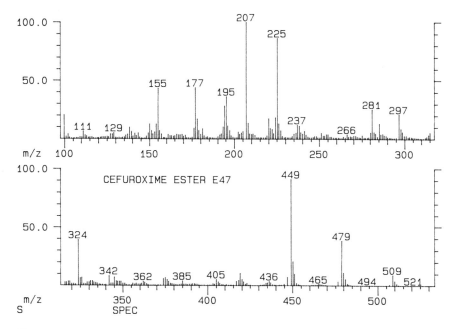

Figure 21 *Negative ion chemical ionization ammonia mass spectrum for the cephalo-sporin ester showing* **M**⁻˙ *at* m/z 509

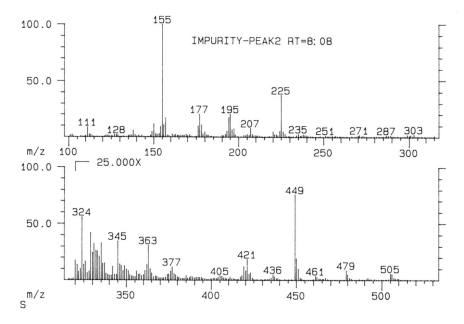

Figure 22 *Negative ion chemical ionization ammonia mass spectrum for the impurity peak at retention time of* 8 *minutes* 8 *seconds in Figure* 20

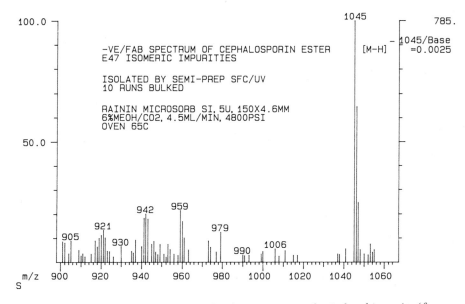

Figure 23 *Negative ion fast atom bombardment spectrum for isolated impurity (from Figure 20)*

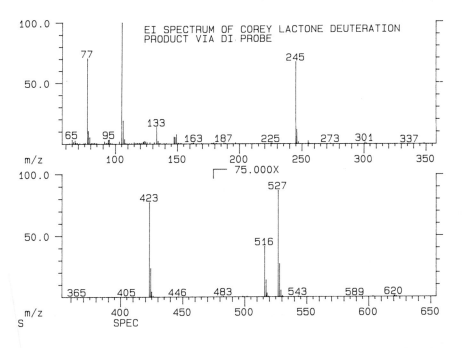

Figure 24 *Direct insertion probe electron impact ionization spectrum of mixed product from Corey lactone showing unaltered isotope pattern for molecular ion region at m/z 516 and strong ion at m/z 527*

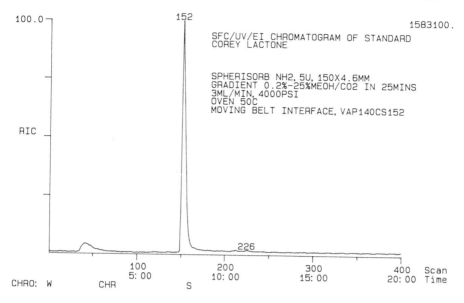

Figure 25 *Total ion current trace for standard Corey lactone*

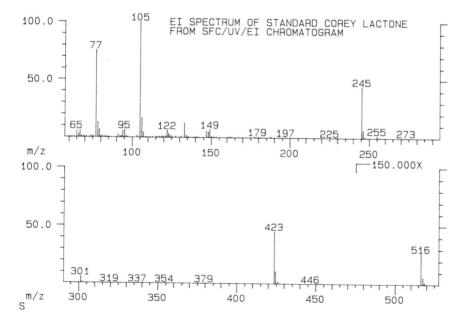

Figure 26 *Electron impact ionization SFC-MS spectrum of Corey lactone*

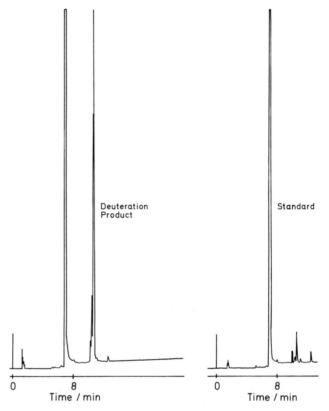

Figure 27 *In-line UV chromatograms for standard and mixed product of Corey lactone. Column: Spherisorb* NH_2, *5 μm, 150 × 4.6 mm; gradient: 0.2%– 25%* CH_3OH/CO_2 *in 25 min, 3 ml min^{-1}; pressure: 4000 psi; oven temperature 50 °C; detector wavelength: 240 nm*

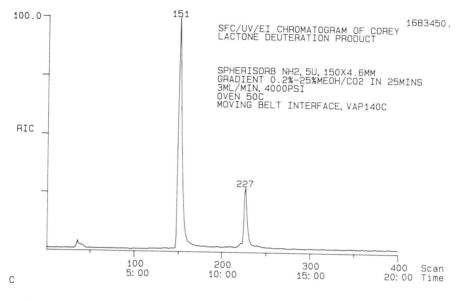

Figure 28 *SFC-MS total ion current trace for mixed product of Corey lactone*

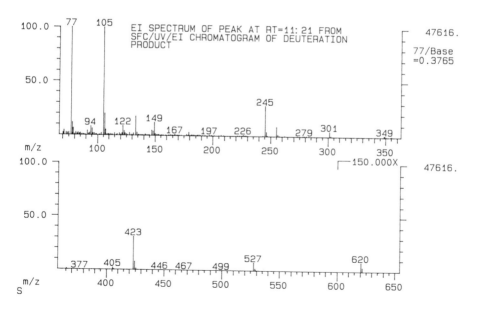

Figure 29 *Electron impact ionization mass spectrum of unknown product*

The erythromycins (5) are antibacterials produced by a strain of *Streptomyces erythreus*[11] and have previously been successfully chromatographed by capillary SFC.[12] Chromatographic methods for analysing this class of compound are of potential interest to the pharmaceutical industry. Commercial erythromycin A was chromatographed by on-line SFC-MS using positive ion chemical ionization. Figure 30 shows the in-line UV trace, and Figure 31 shows the TIC trace. Spectra giving molecular weight information as well as diagnostic fragment ions were obtained for erythromycin A itself, Figure 32, and for minor related impurities Figure 33, Figure 34, which were detected by monitoring the m/z 158 (6) ion as shown in Figure 35.

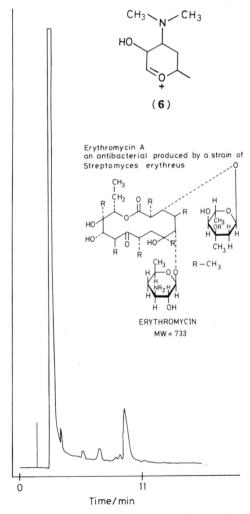

Figure 30 *SFC-UV trace for erythromycin A. Column: Spherisorb* NH_2 *5 μm, 150 × 4.6 mm; mobile phase: 8%* CH_3OH/CO_2, *2.5 ml min*$^{-1}$*; pressure: 4000 psi; oven temperature: 65 °C; detector wavelength: 215 nm*

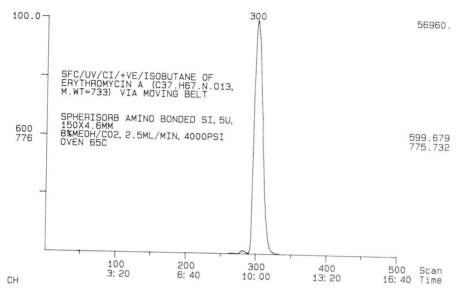

Figure 31 *Total ion current trace for SFC-MS of erythromycin A*

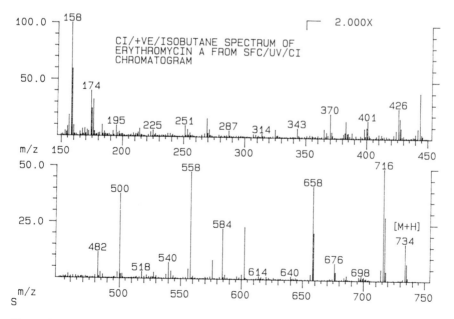

Figure 32 *Positive chemical ionization i-butane mass spectrum for erythromycin A showing* [M + H]⁺ *at m/z 734*

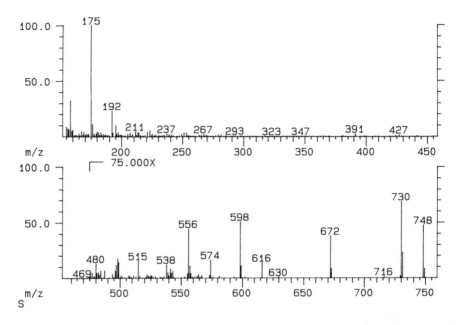

Figure 33 *Positive chemical ionization i-butane mass spectrum of small impurity showing* $[M + H]^+$ *at* m/z 748

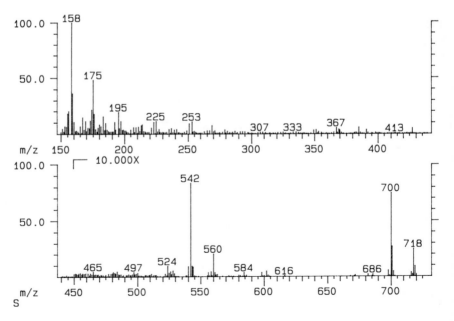

Figure 34 *Positive chemical ionization i-butane mass spectrum of small impurity showing* $[M - H]^+$ *at* m/z 718

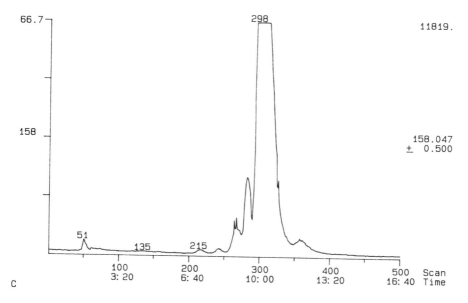

Figure 35 *SIM plot of* m/z *158 for detection of erythromycin A and minor impurities*

4 Conclusion

SFC using packed columns has proved to be a powerful chromatographic technique applicable to a wide range of compounds and problems encountered in the pharmaceutical industry. Interfacing the SFC chromatograph to a MS *via* the moving belt interface is a far less daunting task than the analogous HPLC-MS interfacing, and the combined SFC-MS instrument provides a system of immense potential for the rapid separation and identification of components in complex matrices.

SFC chromatographs can be constructed from available HPLC hardware to produce a system that can easily be reconfigured back to HPLC mode if required.

Packed column SFC using UV detection shows potential for the isolation of clean samples for further off-line MS (*i.e.* FAB) or NMR spectroscopy.

References

1. T. R. Covey, E. D. Lee, A. P. Bruins, and J. D. Henion, *Anal. Chem.*, 1986, **58**, 1451A.
2. M. J. Hayes, E. P. Lankmayer, P. Vouros, and B. L. Karger, *Anal. Chem.*, 1983, **55**, 1752.
3. A. J. Berry, D. E. Games, and J. R. Perkins, *J. Chromatogr.*, 1986, **363**, 147.

4. P. J. Schoenmakers and F. C. C. J. G. Verhoeven, *Trends Anal. Chem.*, 1987, **6**, 10.
5. R. K. Houck, *J. High Resolut. Chromatogr., Chromatogr. Commun.*, 1986, **9**, 4.
6. R. D. Gere, *Science*, 1983, **222**, 253.
7. G. Albers-Schonberg, B. H. Arison, J. C. Chabala, A. W. Douglas, P. Eskola, M. H. Fisher, A. Lusi, H. Mrozik, J. L. Smith, and R. L. Tolman, *J. Am. Chem. Soc.*, 1981, **103**, 4216.
8. H. Mishima, M. Kurabayashi, C. Tamura, S. Sato, H. Kuwano, A. Saito, and A. Aoki, Abstract Papers 18th Symp. Chem. Natural Products, pp. 309–316, Kyoto, 1974.
9. M. Barber, R. Bordoli, R. D. Sedgwick, and A. N. Tyler, *J. Chem. Soc., Chem. Commun.* 1981, **7**, 325.
10. J. Meili and J. Seibl, *Org. Mass Spectrom.*, 1984, **19**, 581.
11. J. R. Martin, R. L. Devault, A. C. Sinclair, R. S. Stanaszek, and P. Johnson, *J. Antibiotics*, 1982, **35**, 426.
12. Lee Scientific 501 SFC Product Literature, 1986.

CHAPTER 8

Fractionation by Coupled Micro-Supercritical Fluid Extraction and Supercritical Fluid Chromatography

M. SAITO*, T. HONDO, and Y. YAMAUCHI

1 Introduction

Although Hannay and Horgarth[1] first found that supercritical fluid had solvating power more than a century ago, two major practical applications of supercritical fluids came into use as recently as the 1960s. These are:

1) Supercritical Fluid Extraction (abbreviated to SFE in this chapter) and,
2) Supercritical Fluid Chromatography (SFC).

Supercritical fluid extraction (SFE), which uses supercritical fluid as the extraction medium, was introduced by Zosel,[2,3] since then, the method has been developed to be an industrial-scale extraction technique as reported by many research groups.[3-10] To obtain a better performance in SFE, the extract must be subjected to separation analysis and the extraction efficiency of the target component checked so that extraction conditions, (*i.e.*, the pressure and temperature of the fluid) can be optimised. For such a purpose, chromatography is an essential technique and various types of chromatography such as thin layer chromatography (TLC), gas chromatography (GC) and high performance liquid chromatography (HPLC) have been employed.

Supercritical fluid chromatography (SFC), which uses a supercritical fluid as the mobile phase, originated with Klesper and co-workers[11,12] in 1962 from high pressure gas chromatography, and was developed by several research groups in the 1960s and 1970s.[13-20] The advances in HPLC instrumentation contributed to the development of sophisticated SFC systems[21-24] in the 1980s, and direct coupling of SFE with SFC was reported separately by the authors[25-28] and Skelton Jr. and co-workers.[29] In the early 1980s, advances in micro-HPLC renewed the interest in SFC. Rapid

203

mass transfer in supercritical mobile phase attracted researchers as it offers high speed separation with high resolution on an open-tubular capillary column, which had not been very successful in liquid chromatography because of slow mass transfer, and also on a packed capillary column. The low fluid consumption encouraged chromatographers to use inflammable and even toxic fluids under high pressure and high temperature. Thus, extensive research work has been carried out by a number of groups.[30−41]

Nowadays, supercritical fluid chromatograhy exists in two major forms:

1) packed-column SFC, and
2) capillary column SFC (open-tubular or packed).

In packed-column SFC, fractionation and collection can be regarded as one of the advantages of SFC by using a fluid that is a gas at atmospheric pressure, such as carbon dioxide. The use of such a mobile phase may allow easy separation of solutes from the column effluent. In 1972, Jentoft and Gouw[17] investigated fractionation in SFC by using a fraction collector enclosed in a high-pressure vessel pressurised with nitrogen gas.

In the last few years, the potential of the separation capability of open-tubular and packed capillary column SFC has been stressed by several research groups.[30−41] However, in capillary column SFC, the fractionation and collection of sample solutes are too difficult or meaningless due to the extremely low sample loading capacity.

Among a variety of supercritical fluids, carbon dioxide is the most widely used, because it is inexpensive, non-toxic, non-flammable, and has comparatively low critical temperature, 31.3 °C, and pressure, 72.9 atm. In SFE, fractionation has been carried out mainly by changing pressure and/or temperature. The use of an adsorbent column has also been proposed in order to isolate components from a mixture, *i.e.*, chromatographic technique. Although present industrial SFE systems are quite different from SFC systems, they have many things in common and their applications overlap each other in one particular area of separation, namely preparative SFC. Table 1 summarises the recent trends in the applications of supercritical fluids.

Table 1 *Recent trends in supercritical fluid applications*

Method	Separation technique	Application
Extraction	Pressure change Temperature change Adsorbent column	Fractionation
		Preparative SFC
Chromatography	Packed column Capillary column	Analytical

2 Instrumentation

At present, various units for pressure measurement *e.g.*, psi, atm, bar, kg cm^{-2}, and Pa are used in the fields of SFE and SFC. In spite of the recommendation to use SI units, Pa is not yet well accepted by many researchers. In this chapter, the authors decided to use kg cm^{-2}, because calculations and apparatus are based on this unit, and it is very close to atm and bar which are used by many people. For the readers' convenience, a unit conversion chart is given in Table 2.

Table 2 *Pressure unit conversion*

	kg cm^{-2}	atm	bar	MPa
kg cm^{-2}	1	0.9678	0.9807	0.09807
atm	1.0332	1	1.0133	0.10133
bar	1.0197	0.9869	1	0.1
MPa	10.197	9.869	10	1

Characteristics of SFE and SFC Systems

In principle, an SFE system consists of a high-pressure pump, an extraction vessel, a back-pressure regulator and a separation vessel. Figure 1 shows schematic diagrams of typical SFE systems. There are three basic types of SFE systems categorized by the different methods which achieve a separation of the extracts from extraction media; system A is based on a pressure reduction which causes a solubility decrease, system B is based on a temperature change which causes a solubility change, and system C is based on the adsorption of the solute. It should be noted that solubility in supercritical fluid is dependent not only on temperature but also on pressure. Therefore, temperature reduction does not always cause a decrease in the solubility of the solute. In practice, a combination of pressure, temperature change, and adsorption techniques are employed for separation of the extract from the fluid.

Figure 2 shows schematic diagrams of HPLC and SFC systems. They are generally very similar to each other but, however, there is a very important device in an SFC system which differentiates this system completely from the HPLC system. This is a back-pressure regulator, which pressurises the system, above the critical pressure of the fluid, from the delivery pump through to the detector, and includes the injector and separation column. Therefore, the main pressure drop takes place abruptly in the back-pressure regulator, whereas on the other hand, in the HPLC system, the pressure drops gradually to atmospheric pressure as the mobile phase solvent passes through the column. In addition to the back-pressure regulator, column heating is also necessary in SFC to keep the mobile phase temperature above the critical temperature of the fluid.

(A)

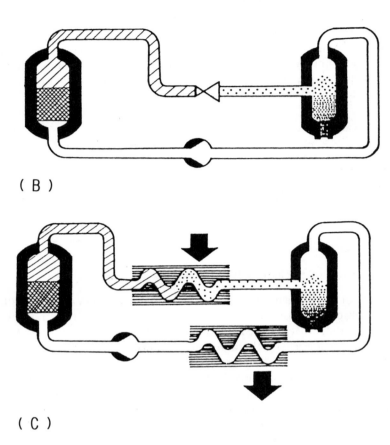

(B)

(C)

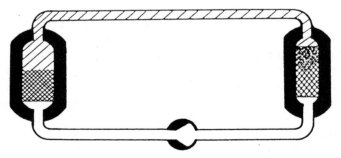

Figure 1 *Schematic diagrams of typical SFE systems. Separation of solutes from the fluid by pressure change* (A), *separation by temperature change* (B), *and separation by adsorbent column* (C). Left: *extraction stage*, right: *separation stage*
(Reproduced by permission from Schneider, Stahl, and Wilke, Verlag Chemie, Weinheim, 1980)[3]

(A)

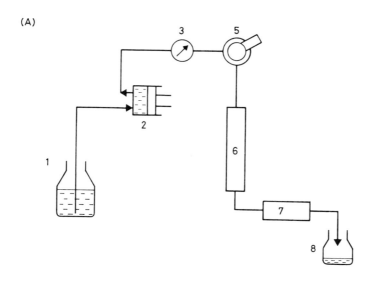

(B)

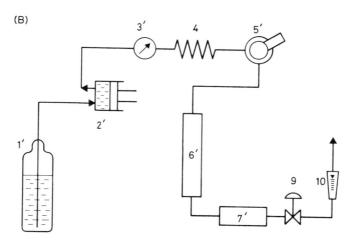

Figure 2 *Schemtic diagrams of HPLC (A) and SFC (B) systems. 1: solvent reservoir, 1': gas cylinder, 2/2': pump, 3/3': pressure meter, 4: heat exchanger, 5/5': injector, 6/6': separation column, 7/7': detector, 8: waste reservoir, 9: back-pressure regulator, 10: gas flow meter*

Back-pressure Device

Conventional Back-pressure Device

As described in the previous section, the device that characterises a super-critical fluid instrument is the back-pressure regulator which is not found in HPLC or solvent extraction systems. There are two different types of back-pressure device currently used in SFE and SFC:

1) simple restrictor, and
2) mechanical and/or electrical feedback regulator.

Table 3 compares these types of the back-pressure devices and their characteristics.

Table 3 *Conventional back-pressure devices*

Type	Dead volume	Applications	Flow dependency
Simple restrictor	$\ll \mu l$	capillary SFC	yes
Feedback regulator	$\gg ml$	packed-column SFC	no

A simple restrictor is a capillary tube having an appropriate length for the required back-pressure and the flow rate.[42-45] This type of back-pressure device is used mainly for open-tubular capillary column SFC. It is easy to make, and has a small dead volume to meet the requirement for use with the column which may have an internal diameter less than 50 μm. However, in order to change the back-pressure, one needs to change the flow rate because the back-pressure is produced only by flow resistance.

The other type of back-pressure device which has been used for SFE and packed column SFC is a mechanical and/or electrical feedback regulator. This type of back-pressure device is a complex regulator which consists of a pressure sensing device and a needle valve.[46] The regulator can control the back-pressure irrespective of the mass flow rate of the fluid. Therefore, it is more convenient than the simple restrictor for the precise examination of retention behaviour in SFC, extraction yield *vs.* pressure in SFE, and other parameters under the same mass flow rate. However, a conventional back-pressure regulator has a very large dead volume, from a few ml to several tens of ml, which would not allow the fractionation of solutes in a fluid from an extraction vessel in micro-SFE, nor from a separation column. This is because the dead volume can be larger than the volume of the extraction vessel or the column. Therefore, the use of such a valve is allowed only when the valve is placed down-stream of the detector and the effluent is not intended to be collected.

Thus, the key to opening the door to fractionation and collection in microscale SFE and preparative SFC is a *low dead volume high-precision back-pressure regulator*. The dead volume must be less than 10 μl, the pressure must be controllable independently of the mass flow rate and by an external pressure reference signal to give the capability for pressure or density programming.

Flow Switching Back-pressure Regulator

In order to meet the requirements described in the previous section, the authors developed a new back-pressure regulator system that is based on a totally different operational principle from a conventional regulator. The regulator valve consists of a needle, which is driven by a solenoid, and valve seat as shown in Figure 3. In general, this type of valve controls the flow resistance of the valve by changing the gap between the valve needle and the seat. The new valve is based on the high speed switching of the fluid flow by periodically opening and closing the flow path. This configuration is very suitable for eliminating the blocking of the valve flow path, which often happens if a conventional type of gap control valve is used. In the new type of valve, precipitated solutes and dry ice originating from carbon dioxide are continually being tapped and forced to pass through the valve.

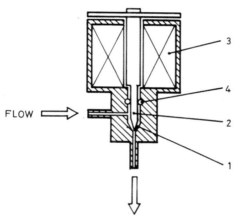

Figure 3 *Cross-sectional view of flow switching back-pressure regulator valve. 1: valve seat, 2: valve needle, 3: needle drive solenoid, 4: needle seal*

In order to regulate the back-pressure at a desired pressure, the valve is used in conjunction with a pressure transducer and control circuitry (Figure 4). A pressure signal from the transducer is fed to the amplifier, and then compared with the pressure set signal. The output of the comparator switches the drive current to the solenoid driving the valve needle to open and close the path periodically.

Figure 5 shows the appearance of the collection device. A glass collection reservoir is connected to the outlet port of the regulator valve by a spring, an arrangement which protects the reservoir from excessive pressure.

The system was connected directly to an HPLC pump pumping methanol at 1 ml min^{-1}. Figure 6 shows the performance of the back-pressure regulator system and gives the pressure recordings at various pressure settings from 50 to 400 kg cm^{-2}. It is seen that the pressure was well regulated within 1% except for small spikes which originated from pump pulsation.

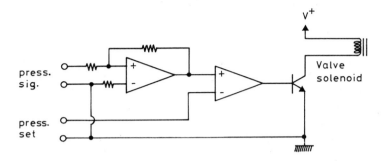

Figure 4 *Block diagram of control circuitry for flow switching back-pressure regulator system*

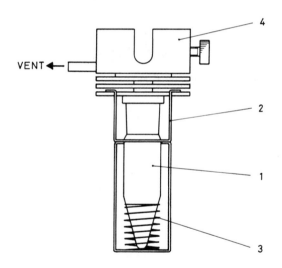

Figure 5 *Configuration of collection device. 1: glass collection reservoir, 2: spring-loaded reservoir retainer, 3: spring, 4: valve adaptor*

Figure 7 shows the pressure recordings of supercritical carbon dioxide at pressure settings from 100 to 350 kg cm^{-2}, at a temperature of 40 °C. Even though the physical properties of supercritical carbon dioxide are quite different from those of methanol; for example, the viscosity is more than ten times lower, the compressibility is extremely high because it is gaseous; in addition, the rapid expansion and cooling of carbon dioxide took place in the valve, and the pressure was also well regulated within 1%. Small pressure dips were due to pump pulsation.

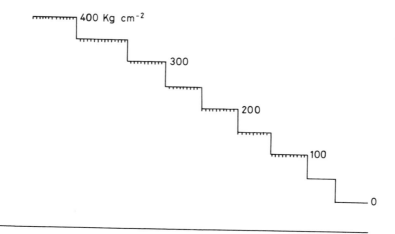

Figure 6 *Performance of back-pressure regulator system with methanol. Conditions are given in text*

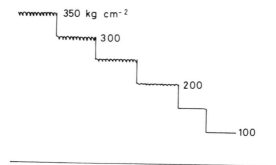

Figure 7 *Performance of back-pressure regulator system with supercritical carbon dioxide. Conditions are given in text*

Extraction Vessel and Separation Column

Safety Precautions

To perform SFE or SFC, an extraction vessel containing the sample or a separation column must be pressurised above the critical pressure of the fluid being used. The extraction vessel and the separation column used in such an experiment are, therefore, classified as high pressure vessels. It should be also noted that any part of the wall of the separation column used in SFC is exposed to the high pressure. Accordingly, unlike in HPLC, even the column outlet-fitting is also under the pressure almost as high as the column inlet pressure and it must withstand the maximum system operating pressure.

In general, the maximum operating pressure of the system must be at least 300 kg cm^{-2}. HPLC components conveniently can be utilized for such a system. However, great care must be taken regarding the pressure resistance of such components. This is because when a gas is compressed to a high pressure at a few hundred kg cm^{-2}, a considerable amount of energy is stored that could be explosively released if the vessel is damaged. On the other hand, when a liquid is compressed, only a little energy is stored due to the lower compressibility. For this reason, safety factors, as specified by the ratio of bursting pressure to maximum operating pressure, which for commercial HPLC components such as columns and tubing are often below 4, are not sufficient for operation with a supercritical fluid phase, which is virtually a high-pressure gas. Therefore, it is strongly recommended to test each component by pressurising it prior to use with water at 1.5 times the maximum operating pressure.

According to design guidelines for high-pressure vessels,[47,48] the minimum wall thickness t is expressed by the following equation based on the safety factor of 4:

$$t \ (\text{mm}) = \frac{P \times \text{i.d.}}{200 \, \eta \, \sigma_a - 1.2 \, P} \qquad \ldots\ldots\ldots (1)$$

where P = maximum operating pressure in kg cm^{-2}, i.d. = internal diameter of vessel or column, σ_a = allowable tensile strength of material in kg mm^{-2}, and η = welding efficiency (η = 1 for seamless construction). For example, assuming that the maximum operating pressure P = 300 kg cm^{-2}, the column or vessel internal diameter i.d. = 4.6 mm and σ_a = 11.3 kg cm^{-2} (allowable tensile strength for 316 stainless steel at 100 °C temperature), the minimum wall thickness t will be:

$$t = \frac{300 \times 4.6}{200 \times 1 \times 11.3 - 1.2 \times 300} = 0.73 \ \text{mm}$$

Accordingly, a 4.6 mm i.d. $\times \frac{1}{4}$ inch o.d. (6.35 mm) column tube will be strong enough for 300 kg cm^{-2} operation.

The wall thickness of the tubing is also expressed by the following equation:[48]

$$t \ (\text{mm}) = \frac{P \times \text{o.d.}}{200 \, \eta \, \sigma_a + 0.8 \, P} \qquad (2)$$

where P = maximum operating pressure in kg cm^{-2}, o.d. = outer diameter of tubing, σ_a = allowable tensile strength of material in kg mm^{-2}, and η = welding efficiency (η = 1 for seamless construction). For example, assuming that the maximum operating pressure P = 300 kg cm^{-2}, the outer diameter of the tubing o.d. = 1.58 mm ($\frac{1}{16}$ inch) and σ_a = 11.3 kg cm^{-2}, the minimum wall thickness t will be:

$$t = \frac{300 \times 1.58}{200 \times 1 \times 11.3 + 0.8 \times 300} = 0.19 \ \text{mm}$$

Thus, the 0.5 mm i.d. $\times \frac{1}{16}$ inch stainless steel tube commonly used in HPLC systems can be used for tubing between high-pressure devices having a maximum operating pressure of 300 kg cm^{-2}.

Extraction Vessel and Separation Column

As was calculated previously, a typical HPLC column having dimensions of 4.6 mm i.d. $\times \frac{1}{4}$ inch (6.35 mm) o.d. has sufficient mechanical strength for the maximum operating pressure of 300 kg cm^{-2}. However, common tubing dimensions of 10 mm i.d. $\times \frac{1}{2}$ inch o.d. (12.7 mm) and 20 mm i.d. $\times$ 1 inch o.d. (25.4 mm) cannot be used in SFC because they both have a limited pressure resistance, 262.6 kg cm^{-2}.

A 4.6 mm i.d. $\times \frac{1}{4}$ inch (6.35 mm) o.d. empty column can be utilized both as a small extraction vessel and a separation column tube. For a vessel or a

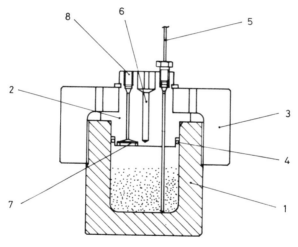

Figure 8 *Cross-sectional view of extraction vessel having* 40 ml *capacity and* 300 kg cm^{-2} *maximum operating pressure. Wall minimum thickness, t: 12.5 mm, i.d.: 37.9 mm. Components: 1: vessel, 2: lid, 3: knurled nut, 4: high durability O-ring, 5: inlet tube, 6: thermocouple port, 7: stainless steel frit, 8: outlet port*

column with an internal diameter greater than 4.6 mm, HPLC components cannot be used because of the insufficient mechanical strength against the gas pressure, and such a vessel or a column must be designed with reference to previous calculations.

Figure 8 shows an example of the extraction vessel having the maximum operating pressure of 300 kg cm^{-2} and the 40 ml capacity, which is large enough for collecting a few ml of a vegetable oil. The lid is securely fastened to the vessel body by the knurled nut, and sealed with a high-durability PTFE O-ring. The nut and lid can be removed by hand without using any special tools. The sample material is placed in the vessel, and supercritical carbon dioxide is introduced through the inlet tube which leads to the bottom of the vessel. Extracts dissolved in carbon dioxide flow out from the outlet port through a stainless steel frit. For the temperature control of the vessel, an aluminium heat jacket is used. When a liquid sample is used, the sample can be stirred during extraction by means of a magnetic stirrer.

Figure 9 shows a different design of a high-pressure vessel with the maximum operating pressure of $300\,kg\,cm^{-2}$ that can be used as an extraction vessel or a separation column.

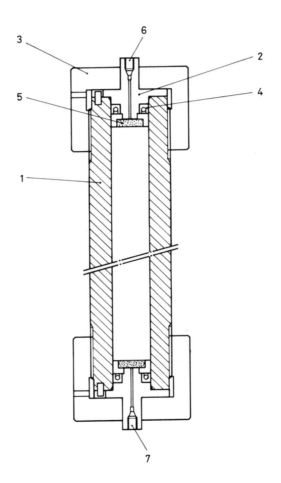

Figure 9 *Cross-sectional view of 20 mm i.d. vessel having $300\,kg\,cm^{-2}$ maximum operating pressure. Wall minimum thickness t: 7.5 mm. It was designed to be used as both an extraction vessel and a separation column. Components: 1: tube, 2: stopper, 3: knurled nut, 4: high-durability PTFE O-ring, 5: stainless steel frit, 6: inlet port, 7: outlet port. Both the inlet and outlet ends have the same configuration*

Detector

In SFE and SFC, various types of detectors including the UV absorption detector, flame ionization detector (FID), mass spectrometer, *etc.*, have been used. Among them, FID is the most popular detector in capillary SFC.[42,43,45,49-51] UV detectors have been the most preferred in packed column SFC.[17-21,23,24,26,27,29,38-41,44] For fractionation, a non-destructive type detector is favourable because it is not necessary to split and waste the effluent containing sample solutes. For this reason, a UV detector is the most feasible among the detectors, which are compatible with supercritical fluids. The UV detector generally offers a stable baseline, high sensitivity, and wide linear dynamic range even with supercritical fluids. In addition, supercritical carbon dioxide is transparent even at 190 nm, which is the short wavelength limit of most of commercial variable wavelength UV detectors. In HPLC, such a short wavelength can only be used as a monitor when pure water is used as the mobile phase, which makes its practical application very limited. In SFE or SFC with supercritical carbon dioxide, even compounds which are generally considered to have little UV absorption, such as lipids, can be detected by a UV detector at a wavelength below 200 nm. In order to make use of this advantage, a variable wavelength or preferably a photo-diode-array multiwavelength UV detector should be used for SFE or SFC. A multiwavelength detector allows UV spectra monitoring as a function of extraction time, or the generation of three-dimensional chromatograms. These data can be stored on a disk and conveniently used for intensive examination after the run by computer technology. In our experiments, which are described in Sections 3 and 4, we used a multiwavelength UV detector.

To utilize any type of a UV detector in SFE and SFC, it must be equipped with a high-pressure flow cell, which can withstand a pressure of at least $300 \, kg \, cm^{-2}$. Metal gaskets of lead or 24 carat gold, and sapphire windows are generally suited for construction of a high-pressure flow cell. However, the UV cut-off wavelength of sapphire is rather high, about 220 nm, which sacrifices the excellent transparency of UV light in supercritical carbon dioxide. Although the tensile strength of quartz is several times lower than that of sapphire and, as a result, the window must be substantially thicker, it is a good material for use when monitoring at a wavelength below 220 nm.

3 Fractionation in Micro-Supercritical Fluid Extraction

It is well-known that wheat germ oil contains tocopherol, which is a fairly expensive material in the pharmaceutical industry. Supercritical carbon dioxide is a suitable extraction medium for vegetable oils. In order to extract tocopherols efficiently from wheat germ powder, one needs to optimise the extraction conditions, *i.e.*, temperature and pressure. For such a purpose, a micro-SFE system, equipped with a multiwavelength UV detector which allows real-time monitoring, and a low dead volume back-pressure regulator, is very effective. We have constructed such a system, incorporating the devices described in Section 2, and examined tocopherol extraction from wheat germ powder.

Experimental

Apparatus

Figure 10 shows a diagram of the hydraulics of a micro-SFE system in which the flow switching back-pressure regulator is utilized. The new system allows efficient fractionation of the extracts owing to the low dead volume, 10 μl, of the regulator. A commercial SFE/SFC system, JASCO model SUPER-100, was used with some modification, including MULTI-320 multiwavelength UV detector.

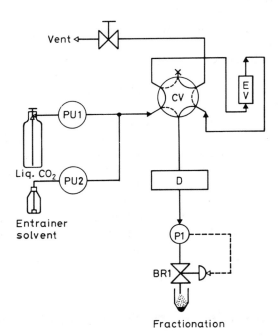

Figure 10 *Diagram of the hydraulics of a micro-SFE system equipped with flow switching back-pressure regulator and multiwavelength UV detector. PU1: carbon dioxide delivery pump, PU2: entrainer solvent delivery pump, EV: extraction vessel, CV: 6-way switching valve, D: UV detector, P1: pressure transducer, BR1: back-pressure regulator valve*

Carbon dioxide is fed to the pump PU1 and delivered to the extraction vessel EV through the six-way switching valve CV, which changes the line to include or by-pass the vessel. Supercritical carbon dioxide containing extracts from the vessel EV is then introduced into the back-pressure regulator valve BR1 *via* the pressure transducer P1. The multiwavelength UV detector D is connected between the switching valve CV and the pressure transducer P1 for real time monitoring of UV spectra to give the extraction profile. As the fluid flows through the back-pressure regulator valve BR1, the fluid pressure is reduced to atmospheric pressure causing the solubility to decrease virtually to zero. The extract in the fluid precipitates and falls from the outlet port of the valve on to which a collecting glass reservoir has been attached.

When the extraction medium is required to have stronger solvating power, the second fluid known as entrainer solvent can be added by the solvent delivery pump PU2.

Materials

Powdered wheat germ was obtained from a grocery store, which was sold as a so-called health food ingredient. Liquefied carbon dioxide used was of an ordinary food additive grade, 99%, in a 30 kg cylinder with a siphon tube.

Results and Discussion

About ten grams of wheat germ were placed in the 40 ml extraction vessel, the detailed configuration of which is given in Figure 8, and extraction was performed with carbon dioxide at various pressures and a constant temperature of 40 °C.

Graph (A) shown in Figure 11 is the extraction profile expressed with UV spectra as a function of extraction time. The extraction pressure was 150 kg cm^{-2}. Graph (B) is the extraction profile expressed as the absorption change at 295 nm, which is about the absorption maximum of tocopherols. According to our previous experiment,[52] only tocopherols exhibit absorption at 295 nm among the components of wheat germ extract. Therefore, this graph is considered to reflect the extraction profile of tocopherols.

Graph (A) in Figure 11 shows that at 150 kg cm^{-2} pressure, lipid components, which have UV absorption only in the wavelength region below 220 nm, are extracted, but only a small amount of tocopherols are extracted. A sharp peak a few minutes after the start of the extraction is not a tocopherol peak but is due to light scattering when the pressure exceeds the critical pressure of carbon dioxide. Another peak at about 8 min is considered to be tocopherols dissolved in the oil, which was swept by the flow of carbon dioxide.

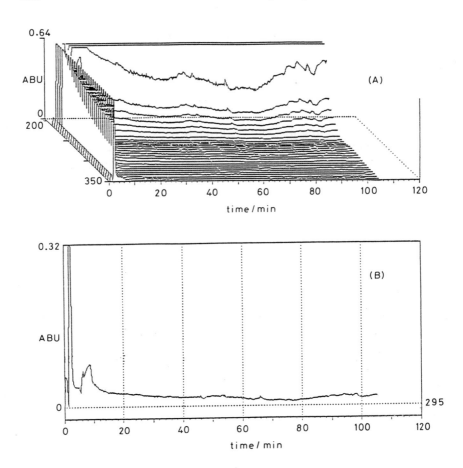

Figure 11 *Extraction profile at* 150 kg cm⁻² *pressure and* 40 °C *temperature. UV spectra* (A), *UV absorption at* 295 nm *as a function of extraction time* (B)

Figure 12 shows extraction profiles at 200 kg cm⁻². The profile at 295 nm, shown in Graph (B), suggests that extraction of tocopherols has started. The first and second peaks appear because of light and tocopherols in oil, as explained above for Figure 11.

Figures 13 (A) and (B) are of extraction profiles at 250 kg cm⁻². Here tocopherols are well extracted and the absorption profile shows a rapid increase, a broad plateau, and then slow decay.

There is little difference in the colours of oils extracted at 150, 200, and 250 kg cm⁻² pressures, but, however, the flavours of the oils could be differentiated easily. The difference in flavours is considered to be due to different constituents of lipid components in the oils extracted at various pressures.

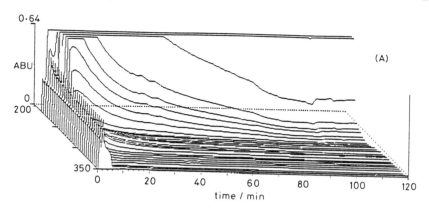

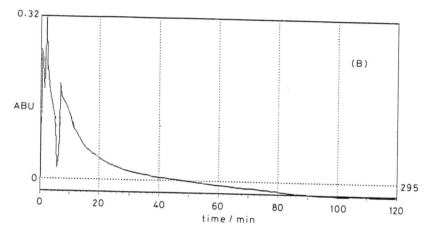

Figure 12 *Extraction profile at* 200 kg cm^{-2}. *Other conditions as for Figure* 11

Table 4 *Pressure* vs. *Extraction Yield*

Compound		Pressure (kg cm^{-2})		
		150	200	250
α-tocopherol	(mg ml^{-1})	0.80	3.30	5.30
	(mg)	0.42	2.31	3.87
β-tocopherol	(mg ml^{-1})	0.20	0.80	1.10
	(mg)	0.11	0.56	0.80
δ-tocopherol	(mg ml^{-1})	—	0.16	0.27
	(mg)	—	0.11	0.20

The extracted amounts of γ-tocopherol were too small to be quantified, and therefore are not included in the table.

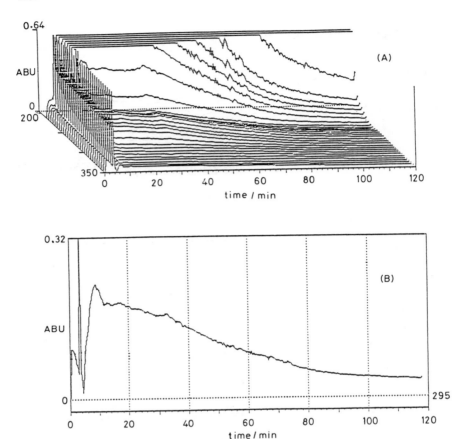

Figure 13 *Extraction profile at* 250 kg cm^{-2}. *Other conditions as for Figure* 11

Figure 14 shows the HPLC chromatogram of the oil extracted at 250 kg cm^{-2}. The peaks were assigned to α-, β-, γ-, and δ-tocopherols with reference to the standard mixture analysis.

Table 4 compares extracted amount of α-, β- and δ-tocopherols at different pressures. The first line of each group shows the tocopherol contents in the extracted oils in terms of mg ml^{-1}, and the second line shows the total extracted amount of tocopherol from 10 g of wheat germ. It can clearly be seen that the extraction yield rapidly increases with increasing the pressure. Below 150 kg cm^{-2}, tocopherols were not well extracted. Therefore, according to our results, the optimum conditions for the extraction of tocopherols from wheat germ powder would be 250 kg cm^{-2} pressure and 40 °C temperature.

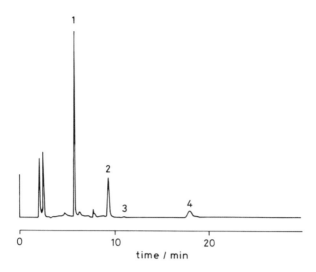

Figure 14 *HPLC chromatogram of extracted oil at* 250 kg cm^{-2} *pressure and* 40 °C *temperature. Peak assignment:* 1: α-tocopherol, 2: β-tocopherol, 3: γ-tocopherol, 4: δ-tocopherol. HPLC conditions: column: JASCO FinePak SIL (4.6 mm i.d. × 250 mm length); mobile phase n-hexane/i-propanol/ acetic acid (99/0.5/0.5) at* 1 ml min^{-1}. *Detection: UV at* 295 nm, 0.08 AUFS

So far, SFE has always been examined for its feasibility as an industrial scale extraction system, however, as we have demonstrated, the micro-SFE system equipped with the newly developed back-pressure control valve successfully extracted several hundred μl of oil from only 10 g of wheat germ. This means that micro-SFE can be used, not only for an optimisation study of extraction conditions for an industrial SFE plant, but it can be used also as a laboratory scale extraction method.

4 Fractionation in Supercritical Fluid Chromatography

In 1985, we examined the analysis of tocopherols from wheat germ by using micro-SFE directly coupled with SFC.[52] However, the system we used was equipped with a conventional back-pressure regulator with a large dead volume; therefore, fractionation was not intended as it was purely an analytical experiment. Figure 15 shows the three-dimensional SFC chromatogram of the SFE extract from wheat germ powder that we obtained in the previous experiment.[52] The extract was introduced into the sample loop of the injector after extraction, then injected onto the separation column. The detailed procedure for sample introduction was described in our previous publication.[27]

Now, we have examined fractionation of tocopherols for wheat germ by using micro-SFE directly coupled with preparative SFC, in which two of the new low dead volume back-pressure regulators were utilized.

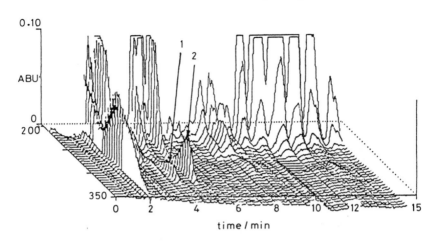

0.10

ABU

0

2 00

350

0   2   4   6   8   10   12   15

time / min

Figure 15 *Three-dimensional SFC chromatogram of SFE extract of wheat germ obtained by analytical SFE/SFC system. Peak assignment: 1: β-tocopherol, 2: α-tocopherol. SFE conditions: sample: wheat germ powder* (400 mg); *pressure:* 300 kg cm^{-2}; *temperature:* 40 °C; *extraction time:* 5 min. *SFC conditions: column: JASCO FinePak SIL* C$_{18}$ (4·6 mm i.d. × 150 mm *length*); *mobile phase:* CO$_2$; *pressure:* 200 kg cm^{-2}; *flow rate* 2 l min^{-1} *as gas at atmospheric pressure*

Experimental

Apparatus

Figure 16 shows a diagram of the hydraulics of the coupled micro-SFE/ preparative SFC system. The SFE section of the system is very similar to that shown in Figure 10 in Section 3, *i.e.*, the preparative SFC section is connected directly to the back-pressure regulator of the SFE so that they are incorporated into a double-stage separation system.

The preparative SFC section consists of a pre-column PC, a separation column C, the same multiwavelength UV detector D, and back-pressure regulator system with the collection reservoir (as in Section 3). First, SFE is performed with carbon dioxide applying the back-pressure by the regulator BR1, while the BR2 is set to zero pressure, in other words, full-open. The extract is adsorbed onto the pre-column which is connected directly to the outlet port of the valve. Therefore, at this stage, carbon dioxide gas flows through the pre- and separation columns. On completion of SFE, the extraction vessel is by-passed by switching the valve. Then, the BR1 is set to zero pressure, to full-open, and the BR2 to the necessary pressure for chromatographic elution. Then, carbon dioxide containing ethanol as a modifier starts eluting the extract adsorbed on the pre-column.

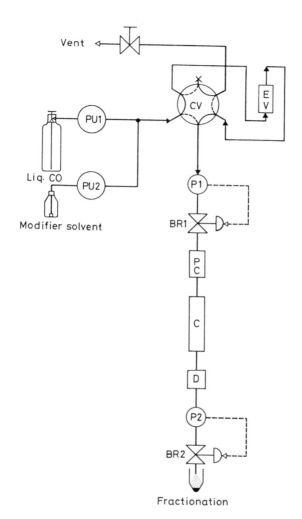

Figure 16 *Hydraulic diagram of coupled micro-SFE/preparative-SFE system.* PU1: *carbon dioxide delivery pump,* PU2: *solvent delivery pump,* CV: *6-way switching valve,* EV: *extraction vessel,* D: *UV detector,* P1: *pressure transducer,* BR1: *back-pressure regulator valve,* PC: *pre-column for adsorption of extracts,* C: *separation column,* D: *UV detector,* P2: *pressure transducer,* BR2: *back-pressure regulator valve*

Materials

Wheat germ used for this experiment was also obtained from a grocery store, but the germ was not powdered, and each grain was a thin disc of about 1–1.5 mm in diameter.

Results and Discussion

Five grams of wheat germ were placed in the extraction vessel of 10 ml volume (8 mm i.d. × 200 mm length) and extraction was performed at 250 kg cm^{-2} pressure and 40 °C temperature for one hour, which were suitable extraction conditions confirmed by the experiment in Section 3. Then, preparative SFC was performed.

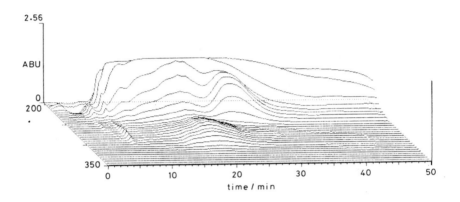

Figure 17 *Three-dimensional SFC chromatogram obtained by micro-SFE/preparative-SFE system. For conditions, see text*

Figure 17 shows the three-dimensional chromatograms of the wheat germ extract obtained by the above procedure. SFC conditions were: back-pressure: 250 kg cm^{-2}; carbon dioxide flow rate: 8 ml min^{-1} as liquid at −5 °C; and modifier: methanol at 0.2 ml min^{-1} flow rate. The separation column used was 10.2 mm i.d. × 250 mm length packed with Develosil silica gel 30 to 50 μm. Silica gel was used instead of silica-ODS, which was used for the analytical work shown in Figure 15, because silica gel generally offers a higher sample loading capacity than silica-ODS. It should be noted that silica-ODS with supercritical carbon dioxide systems does not function as a reversed-phase system, but, depending on sample solutes, often functions as a normal-phase system with weaker stationary phase properties than silica gel itself. Therefore, similar chromatograms could be obtained to some extent by using an appropriate amount of polar modifier.

However, the chromatographic peaks (Figure 17), are very broad and different from those obtained from an analytical column (Figure 15). This means that the column efficiency was very low, which may be due to the larger particle diameter of the packing material. Also lipid components besides tocopherols were considered to have been overloaded.

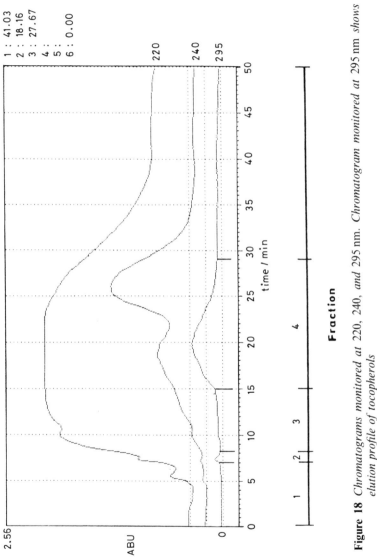

Figure 18 *Chromatograms monitored at 220, 240, and 295 nm. Chromatogram monitored at 295 nm shows elution profile of tocopherols*

Chromatograms monitored at 220, 240, and 295 nm are shown in Figure 18. These were selected from the stored three-dimensional data from the run, shown as a chromatogram in Figure 17. As mentioned in Section 3, the chromatogram at 295 nm represents the tocopherol elution profile. The numbers under the time axis show the time frames and corresponding fractions collected with reference to real time monitoring of the chromatogram. These chromatograms suggested that fraction 4 contained tocopherols.

There were clear differences in the colours of the fractions, and fractions 3, 4, and 5 showed two liquid layers. Ethanol, which was added as a modifier was in the upper layer, and the fractionated oil was in the lower layer in each tube, showing that these oils have solubilities too low for them to be dissolved completely in ethanol.

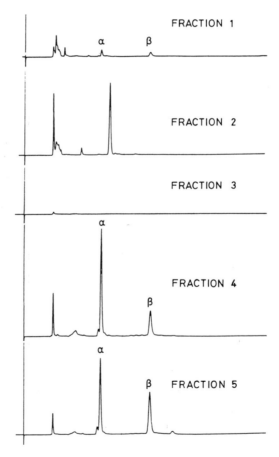

Figure 19 *HPLC chromatograms of collected oils. HPLC conditions: Column: JASCO FinePak SIL (4.6 mm i.d. × 250 mm length), mobile phase: n-hexane/isopropanol/acetic acid (99/0.5/0.5) at 1 ml min⁻¹; detection: UV at 295 nm, 0.08 AUFS, sample: 100 μg each of fractionated oil*

On the other hand, fractions 1 and 2 did not exhibit such layers, maybe due to smaller quantities of collected oils and/or higher solubilities in ethanol.

Figure 19 shows HPLC chromatograms of the fractions 1–5. As expected, fractions 1 and 2 contained only small amounts of tocopherols, while fractions 4 and 5 contained higher amounts of tocopherols. Time frame 5 showed comparatively low absorption at 295 nm, however, it could be recognized as a broad peak or a tailing of the peak in time frame 4. It should be noted that fraction 5 contained more β-tocopherol than fraction 4. This suggests that the peak in time frame 4 contained more α- and less β-tocopherols in the earlier part, and *vice versa* in the later part.

Table 5 *Collected amounts of tocopherols and oil*

Fraction	Collected oil (mg)	Tocopherols (μg)	
		α	β
1	1.1	0.3	0.3
2	1.9	0.1	—
3	82.5	—	—
4	115.4	480.0	100.0
5	11.3	68.8	68.8
Total	212.2	549.2	169.1

$$\text{Recovery} = \frac{\text{Amount collected}}{\text{Amount adsorbed}} \times 100\%$$

$$= \frac{212.2 \text{ mg}}{220.0 \text{ mg}} \times 100\% = 96.5\%$$

Table 5 lists the amounts of collected tocopherol and oil for each fraction. The total recovery was 96.5% in terms of the sum of fractionated oil divided by the amount adsorbed onto the pre-column, measured with a balance. This recovery means that the fractionation was carried out successfully without significant loss of the solutes even though the solutes were sprayed out from the back-pressure regulator.

However, enrichment of tocopherols was not very successful. Fraction 4 contained only 480 μg of α- and 100 μg of β-tocopherols in 115.4 mg of collected oil. These are 0.42 and 0.087% by weight in the fractionated oil, while the percentage content of α-tocopherol in the extracted oil without SFC separation was 0.31%. It is partly because the time frame 4 was a little too wide and the low-concentration portion of the peak was included. But the main cause may be attributed to the poor efficiency of the separation column.

Although the efficiency is not totally satisfactory, it has been proven that double-stage micro-SFE/preparative SFC is possible and fractionation can be carried out without difficulty in a similar manner to HPLC.

5 Conclusion

In order to obtain better results in coupled SFE/preparative SFC, we still have a lot of work to be carried out, such as making suitable packing materials, optimisation of linear velocity, increasing maximum sample loading capacity, *etc.* A recycling technique in high-pressure system, though it is not easy to accomplish, might be very powerful for preparative SFC. The effective column length can be extended by the factor of the number of recycles, while the pressure drop across the column is kept to a minimum, which is desirable in SFC to maintain the column efficiency maximum. Discarding unnecessary components during a recycling operation, which is commonly performed in preparative HPLC, and recycling of only those peak components containing the target compound may reduce the sample loading capacity required.

In conclusion, the authors would like to emphasise that applications of supercritical fluids are not limited only to analytical SFC. One must respect the advantage of supercritical carbon dioxide over organic solvents; that is, the easy separation of solutes at low temperatures and in an oxygen-free environment.

Acknowledgement

The authors wish to thank Messrs. M. Sugawara, H. Kashiwazaki and H. Konishi for their assistance in the development of the micro-SFE/preparative-SFC system.

References

1. J. B. Hannay and J. Hogarth, *Proc. Roy. Soc.* (*London*), 1879, **29**, 324.
2. K. Zosel, Austrian Patent Appl. 16. 4. 1963.
3. 'Extraction with Supercritical Gases', *eds.* G. M. Schneider, E. Stahl, and G. Wilke, Verlag Chemie, Weinheim, 1980.
4. E. Stahl, E. Schutz, and H. Mangold, *J. Agric. Food Chem.*, 1980, **28**, 1153.
5. E. Stahl and E. Schutz, *Planta Med.*, 1980, **40**, 262.
6. H. Coenen and P. Rinza, 'Tech. Mitt. Krupp-Werksberichte', 1981, **39**, H1, Z1.
7. *Chem. Ind. (London)*, 19 June 1982.
8. G. Brunner and S. Peter, *Ger. Chem. Eng.*, 1982, **5**, 181.
9. 'Supercritical Fluid Technology', *eds.* J. M. L. Penninger, M. Randosz, M. A. McHugh, and V. J. Krukonis, Elsevier, Amsterdam, 1985.

10. M. A. McHugh and V. J. Krukonis, 'Supercritical Fluid Extraction', Butterworths, Boston, 1986.
11. E. Klesper, A. H. Corwin, and D. A. Turner, *J. Org. Chem.*, 1962, **27**, 700.
12. N. M. Karayannis, A. H. Corwin, E. W. Baker, E. Klesper, and J. A. Walter, *Anal. Chem.*, 1968, **40**, 1736.
13. M. N. Myers and J. C. Giddings, *Anal. Chem.*, 1966, **38**, 294.
14. M. N. Myers and J. C. Giddings, *Anal. Chem.*, 1965, **37**, 1453.
15. J. C. Giddings, M. N. Myers, and J. W. King, *J. Chromatogr. Sci.*, 1969, **7**, 276.
16. R. E. Jentoft and T. H. Gouw, *J. Chromatogr. Sci.*, 1970, **8**, 138.
17. R. E. Jentoft and T. H. Gouw, *Anal. Chem.*, 1972, **44**, 681.
18. T. H. Gouw and R. E. Jentoft, *J. Chromatogr.*, 1972, **68**, 303.
19. M. Novotny, W. Bertsch, and A. Zlatkis, *J. Chromatogr.*, 1971, **61**, 17.
20. D. Bartmann and G. M. Schneider, *J. Chromatogr.*, 1973, **83**, 135.
21. D. R. Gere, R. Board, and D. McManigill, *Anal. Chem.*, 1982, **54**, 736.
22. T. Greibrokk, A. L. Blilie, E. J. Johansen, and E. Lundanes, *Anal. Chem.*, 1984, **56**, 2681.
23. K. Jinno, M. Saito, T. Hondo, and M. Senda, *Chromatographia*, 1986, **21**, 219.
24. K. Jinno, T. Hoshino, T. Hondo, M. Saito, and M. Senda, *Anal. Lett.*, 1986, **19**, 1001.
25. K. Sugiyama, M. Saito, and A. Wada, Japanese Pat. Appl. No. 58–117773 (1984).
26. M. Saito, K. Sugiyama, T. Hondo, M. Senda, and S. Tohei, International HPLC Symposium Kyoto, Jan., 1985, Abstract p. 84.
27. K. Sugiyama, M. Saito, T. Hondo, and M. Senda, *J. Chromatogr.*, 1985, **332**, 107.
28. K. Sugiyama, M. Saito, and A. Wada, U.S. Pat. No. 4,597,943 (1986).
29. R. J. Skelton, Jr., C C. Johnson, and L. T. Taylor, *Chromatographia*, 1986, **21**, 4.
30. M. Novotny, S. R. Springston, P. A. Peaden, J. C. Fjeldsted, and M. L. Lee, *Anal. Chem.*, 1981, **53**, 407A.
31. P. A. Peaden, J. C. Fjeldsted, M. L. Lee, S. R. Springston, and M. Novotny, *Anal. Chem.*, 1982, **54**, 1090.
32. R. D. Smith, W. D. Felix, J. C. Fjeldsted. and M. L. Lee, *Anal. Chem.*, 1982, **54**, 1883.
33. P. A. Peaden and M. L. Lee, *J. Liq. Chromatogr.*, 1982, **5**, 179.
34. P. A. Peaden and M. L. Lee, *J. Chromatogr.*, 1983, **259**, 1.
35. S. R. Springston and M. Novotny, *Anal. Chem.*, 1984, **56**, 1762.
36. J. C. Fjeldsted and M. L. Lee, *Anal. Chem.*, 1984, **56**, 619A.
37. R. D. Smith, H. T. Kalinoski, H. R. Udseth, and B. W. Wright, *Anal. Chem.*, 1984, **56**, 2476.
38. Y. Hirata and F. Nakata, *J. Chromatogr.*, 1984, **295**, 315.
39. T. Takeuchi, D. Ishii, M. Saito, and K. Hibi, *J. Chromatogr.*, 1984, **295**, 323.

40. Y. Hirata, *J. Chromatogr.*, 1984, **315**, 31.
41. Y. Hirata, *J. Chromatogr.*, 1984, **315**, 39.
42. M. L. Lee and K. E. Markides, Proceedings of the Seventh International Symposium on Capillary Chromatography, p. 575, Gifu, Japan, May, 1986.
43. M. V. Novotny and P. A. David, Proceedings of the Seventh International Symposium on Capillary Chromatography, p. 586, Gifu, Japan, May, 1986.
44. Y. Hirata, F. Nakata, and M. Kawasaki, Proceecnatio of the Seventh International Symposium on Capillary Chromatography, p. 598, Gifu, Japan, May, 1986.
45. H. E. Schwartz, R. G. Brownlee, and E. J. Guthrie, Proceedings of the Seventh International Symposium on Capillary Chromatography, p. 637, Gifu, Japan, May, 1986.
46. Technical Brochure for Pressure Regulators, Tescom Coporation, Pressure Control Division, MN, U.S.A.
47. Japanese Industrial Standard, Alternative Standard for Construction of Pressure Vessels, JIS B 8250-1983, Japanese Standards Association, Tokyo 107, Japan.
48. Japanese Regulations of High Pressure Gases, 17th Edn., 1985, High Pressure Gas Association, Tokyo 105, Japan.
49. M. G. Rawdon, *Anal. Chem.*, 1984, **56**, 831.
50. T. A. Norris and M. G. Rawdon, *Anal. Chem.*, 1984, **56**, 1767.
51. T. L. Chester, *J. Chromatogr.*, 1984, **299**, 424.
52. T. Hondo, M. Saito, and M. Senda, Pittsburgh Conference Abstracts, No. 484 (1986).

Subject Index

Compounds used as supercritical fluids or modifiers are indexed separately unless properties or specific applications are recorded. Individual compounds which have been separated by SFC are listed in the Compound Index.

Compound Index

For a summary of the types of compounds analysed by SFC see Table 4, p. 48. Page numbers in italic indicate compounds depicted in chromatograms.

Compounds used as mobile phases or modifiers in supercritical fluid chromatography are shown in italic. They are listed only when their properties are discussed or they are used for specialised applications.